T0100619

FUNDAMENTALS OF FLUVIAL GEOMORPHOLOGY

Fundamentals of Fluvial Geomorphology provides an accessible introduction to the study of river landforms and the processes that shape them. Rivers are significant geomorphological agents which show an amazing diversity of form and behaviour, reflecting the wide range of environments in which they are found. Highly dynamic in nature, river channels adjust and evolve over timescales that range from seconds to tens of thousands of years. This book examines how river systems operate and respond to change and why this understanding is needed for successful river management.

The book provides a coherent overview of the main concepts in fluvial geomorphology, together with recent developments in river channel management, clearly illustrating why an understanding of fluvial geomorphology is vital in channel preservation, environmentally sensitive design and the restoration of degraded river channels. Starting with an introduction to the fluvial system, the book moves on to cover:

- Flow and sediment regimes: flow generation; flow regimes; sediment sources, transfer and yield.
- Channel process: flow characteristics; processes of erosion and sediment transport; interactions between flow and the channel boundary; deposition.
- Channel form and behaviour: controls on channel form; channel adjustments; floodplain development; form and behaviour of alluvial and bedrock-influenced channels.
- Response to change: how the system responds to change; reconstructing past changes; impacts of climate change, human activity, tectonics and base-level change.
- River management: the fluvial hydrosystem; environmental degradation; environmentally sensitive engineering techniques; river restoration; the role of the fluvial geomorphologist.

Fundamentals of Fluvial Geomorphology is an indispensable text for undergraduate students. It provides straightforward explanations for important concepts and mathematical formulae, backed up with conceptual diagrams and appropriate examples from around the world to show what they actually mean and why they are important. A colour plate section also shows spectacular examples of fluvial diversity.

Ro Charlton lectures in the Geography Department at the National University of Ireland, Maynooth, where she teaches undergraduate and postgraduate courses in fluvial geomorphology and hydrology. Her research interests include meandering river canyons in caves and the impact of climate change on water resources.

FUNDAMENTALS OF FLUVIAL GEOMORPHOLOGY

Ro Charlton

Routledge
Taylor & Francis Group

LONDON AND NEW YORK

First published 2008 by Routledge
2 Park Square, Milton Park, Abingdon, Oxon, OX14 4RN

Simultaneously published in the USA and Canada
by Routledge
711 Third Avenue, New York, NY 10017

Routledge is an imprint of the Taylor & Francis Group, an informa business

©2008 Rosemary Charlton
Typeset in Garamond by Keyword Group Ltd
Printed and bound in Great Britain by the MPG Books Group

All rights reserved. No part of this book may be reprinted or reproduced or
utilised in any form or by any electronic, mechanical, or other means, now
known or hereafter invented, including photocopying and recording,
or in any information storage or retrieval system, without permission
in writing from the publishers.

British Library Cataloguing in Publication Data
A catalogue record for this book is available from the British Library

Library of Congress Cataloging in Publication Data
Charlton, Ro.
Fundamentals of fluvial geomorphology/Ro Charlton.
p. cm.
Includes bibliographical references.
1. Fluvial geomorphology. I. Title.
GB561.C475 2007
551.3′55—dc22
2007014030

ISBN10: 0-415-33453-5 (hbk)
ISBN10: 0-415-33454-3 (pbk)
ISBN10: 0-203-37108-9 (ebk)

ISBN13: 978-0-415-33453-2 (hbk)
ISBN13: 978-0-415-33454-9 (pbk)
ISBN13: 978-0-203-37108-4 (ebk)

To my parents

CONTENTS

LIST OF FIGURES

LIST OF BOXES

LIST OF PLATES

LIST OF COLOUR PLATES

AIMS AND SCOPE OF THE BOOK

This book aims to provide an accessible introduction to the subject of fluvial geomorphology for junior undergraduate students of geography and related disciplines. In addition, it will provide supporting material for more senior undergraduates who are looking for basic explanations to aid in the understanding of advanced texts and journal articles. Since fluvial geomorphology is a rapidly expanding, multi-disciplinary area, the book will also be of interest to those who require a broad overview of the subject, such as ecologists, geologists and engineers.

Although introductory physical geography and geomorphology textbooks include chapters on fluvial geomorphology, the inevitable constraints imposed by the scope of such texts allows only relatively brief coverage. On the other hand, most of the fluvial geomorphology texts that are currently available are pitched at third-year level and beyond. Students at a more junior level may struggle to understand the underlying concepts, especially those who have a limited background in mathematics and science. This book bridges that gap. It has been written to provide clear, straightforward explanations of concepts and formulae, and little previous knowledge of maths and science is assumed. The material is organised to build on itself throughout the book, with the development of concepts and ideas that have been introduced in earlier chapters. These are illustrated with case studies and examples to develop the student's interest. More difficult concepts and additional formulae have been placed in self-contained 'boxes', which can be bypassed if desired. Boxes are also included in some chapters to illustrate field techniques and selected case studies. New terms are introduced using bold text and many of these are included in an extensive glossary at the end of the book. In recommending further reading, it is assumed that readers will progress initially to more advanced textbooks for further information, although selected journal articles have also been listed to include illustrative case studies and some 'classics'. Useful web sites are also included. The owners of faculty sites have been contacted to confirm the stability of the site in the foreseeable future and to request that a forwarding link is provided if necessary.

Fluvial geomorphology is interpreted in a broad sense in this book, which includes chapters on hydrological and hillslope processes as well as large-scale sediment transfer. This is felt to be appropriate for an introductory text, since an appreciation of the interactions between hillslopes and channel–floodplain systems is necessary to understand how the fluvial system responds to change. A holistic, catchment-wide approach is also widely advocated for successful river channel management.

There are ten chapters, starting with an overview of some of the key concepts of fluvial geomorphology in Chapter 1, which introduces the huge variety of fluvial forms that are seen worldwide. Following on from this,

Chapter 2 discusses the fluvial system, examining equilibrium, scale and the complex interrelationships that exist between variables. Chapter 3 focuses on relevant aspects of hydrology, with particular emphasis on the characteristics of different flow regimes, flood frequency–magnitude relationships and channel-forming flows. A general discussion of sediment sources is provided in Chapter 4, which covers weathering, mass wasting and slope erosion processes. Chapter 5 goes on to explain how sediment is transferred through the fluvial system. The importance of sediment storage is highlighted, with an examination of scale effects and the historical legacy of past events, such as glaciations and human activity. Flow processes within the channel are considered in Chapter 6, which provides a basic overview of the properties of fluid flow, flow resistance and boundary layers. Chapter 7 discusses processes of erosion, sediment transport and deposition, highlighting the differences between bedload and suspended load transport. There is also a section on erosion in bedrock channels, which have been neglected by many previous texts. Channel form is examined in Chapter 8, which starts by considering the various controls on morphology, the nature of morphological adjustments and the space and time scales over which they take place. A wide range of alluvial and bedrock-influenced channels are included and relationships between floodplain morphology and channel processes are also examined. Following on from this, Chapter 9 considers the response of fluvial systems to environmental change. Approaches to reconstructing and understanding the nature of past changes are outlined, followed by a discussion of the ways in which river systems have been influenced by climate change, human activity, tectonics and changes in base level. The focus of Chapter 10 is on management, starting with the management problems that are associated with river channels and the reasons why channel engineering is carried out. Some of these management problems have been discussed in preceding chapters and the student is referred back to the relevant sections. Traditional and more recently developed environmental engineering techniques are examined, and the final section focuses on river restoration, providing an overview of the main techniques used and the considerations that need to be taken into account if these are to be successful.

ACKNOWLEDGEMENTS

There are a number of people who I would like to thank for their contribution to this book, not least the many friends and colleagues who have provided support and encouragement during its preparation. Special mention must go to Jim Keenan for his patience and skill in reproducing the artwork. I am very grateful to the three reviewers of the first draft, who provided helpful and constructive criticisms, which I have tried to address in the final version. The comments and suggestions made by the reviewers of the initial proposal were also very helpful. Paddy Charlton and Allan Hale read through the first draft and, as non-specialists, commented on the clarity and accessibility of the text. They also corrected numerous typos and grammatical errors. Thanks are also due to Ian Benson for several interesting discussions on the nature of fluid mechanics and sediment transport and for his comments on the final versions of Chapters 6 and 7. Bernadette Gardiner and the library staff at the National University of Ireland, Maynooth, were always helpful and efficient in dealing with numerous inter-library loans. Some of the material in this book was class-tested on students in the Department of Geography, NUI, Maynooth, who provided helpful feedback. The NUI, Maynooth, Publications Committee awarded a grant towards the cost of photographs and other illustrations.

The copyright of the photographs remain held by the individuals who kindly supplied them. The author and publishers would like to thank the following for granting permission to reproduce material in this work:

Airphoto – Jim Wark for permission to reproduce Plate 10.2 and Colour Plates 4, 5, 9, 10, 11, 15, 17 and 18.

American Society of Civil Engineers, Reston VA, for permission to reproduce Figure 9.5(a), after figure 8.7 in D.J. Harbor, S.A. Schumm and M.D. Harvey, Tectonic control of the Indus River in Sindh, Pakistan, in S.S. Schumm and B.R. Winkley (eds), *The Variability of Large Alluvial Rivers* (ASCE Press, New York, 1994).

Blackwell Publishing, Oxford, for permission to reproduce Figure 8.4(a), after figure 6.3 in M. Church, Channel Morphology and Typology, in P. Calow and G.E. Petts, *The Rivers Handbook* (Blackwell Scientific, London, 1992), and Figure 8.2, after figure 1 in P.J. Ashworth and R.I. Ferguson, Interrelationships of channel processes, changes and sediments in a proglacial river, *Geografiska Annaler*, 68 (1986), 361-71, and Figure 8.10, after figure 4.9 in G. Brierley and K. Fryirs, *Geomorphology and River Management: Applications of the River Styles Framework* (Blackwell Science, London, 2005).

Rhett Butler, WildMadagascar.org for permission to reproduce Plate 4.3.

Cambridge University Press, Cambridge, for permission to reproduce Figure 6.10, after figure 1 in K. Shiono and D.W. Knight, Turbulent open channel flows with variable depth across the channel, *Journal of Fluid Mechanics*, 222 (1991), pp. 617–46.

Canadian Society of Petroleum Geologists, Calgary, for permission to reproduce Figure 7.6, after figure 2 in A.C. Brayshaw, The Characteristics and origin of Cluster Bedforms in Coarse-grained Alluvial Channels, in C.H. Koster and R.H. Stell, (eds), *Sedimentology of Gravels and Conglomerates*, Canadian Society of Petroleum Geologists Memoir 10 (1984), pp. 77–85.

University of Chicago Press, Chicago, for permission to reproduce Figure 5.6, after figure 6 in J.D. Milliman and P.M. Syvitski, Geomorphic/tectonic control of sediment discharge to the ocean: the importance of small mountainous rivers, *Journal of Geology*, 100 (1992), pp. 525–44.

Elsevier, London, for permission to reproduce Figure 8.4(b), after figure 3 in T.J. Coulthard, J. Lewin and M.G. Macklin, Modelling differential catchment response to environmental change, *Geomorphology*, 69 (2005), pp. 222–41, copyright 2005 Elsevier.

Foerderkreis "Rettet die Elbe", Hamburg (Germany) for permission to reproduce Colour Plate 16.

Connell Foley for permission to reproduce Colour Plate 14.

Geological Society of America, Boulder CO, for permission to reproduce Figure 8.13, after figure 4 in E.A. Keller, Development of alluvial stream channels: a five-stage model, *Geological Society of America Bulletin*, 83 (1972), pp. 1531–6.

Hodder Education, London, for permission to reproduce Figures 6.3, 8.5, 8.8(c)–(d) and 8.15, after figures 4.1B, 5.3, 5.13 and 5.23 in D. Knighton, *Fluvial Forms and Processes: A New Perspective* (Arnold, London, 1998), and Figure 6.8(a), after figure 2.2 in A. Robert, *River Processes: An Introduction to Fluvial Dynamics* (Arnold, London, 2003).

Helen Houghton-Carr for permission to reproduce Plate 7.1.

Tim McCabe, *National Resources Conservation Service* for permission to reproduce Plate 8.1.

Jeanne Meldon for permission to reproduce Plate 6.1.

John Wiley & Sons, Chichester, for permission to reproduce Figure 2.4, after figure 12.1 in S.A. Schumm, Variability of the Fluvial System in Space and Time, in T. Rosswell, R.G. Woodmansee and P.G. Risser (eds), *Spatial and Temporal Variability in Biospheric and Geospheric Processes* (Wiley, Chichester, 1988), pp. 225–50; Figure 3.3(b) and 3.4, after figures 10.2 and 10.11 in D.A. Knighton and G.D. Nanson, Distinctiveness, Diversity and Uniqueness in Arid Zone River Systems, D.S.G. Thomas (ed.), *Arid Zone Geomorphology* (Wiley, Chichester, 1997), pp. 185–203; Figure 7.7, after figure 5.10A and C, in I. Reid, J.C. Bathurst, P.A. Carling, D.E. Walling and B.W. Webb, Sediment Erosion, Transport and Deposition, in C.R. Thorne, R.D. Hey and M.D. Newson (eds), *Applied Fluvial Geomorphology for River Engineering and Management* (Wiley, Chichester, 1997), pp. 95–135; Figure 8.14, after figure 22.2 in A.J. Markham and C.R. Thorne, Geomorphology of Gravel-bed River Bends, in P. Billi, R.D. Hey, C.R. Thorne and P. Taconi (eds), *Dynamics of Gravel-bed Rivers* (Wiley, Chichester, 1992), pp. 433–50, and Figure 9.4, after figure 7.6, in N.D. Gordon, T.A. McMahon, B.L. Finlayson, C.J. Gippel and R.J. Nathan, *Stream Hydrology: An Introduction for Ecologists*, second edition (Wiley, Chichester, 2004).

Gerald Nanson for permission to reproduce Plate 8.2 and Colour Plate 8.

NASA for permission to reproduce Colour Plates 6 and 12.

Oxford University Press, Oxford, for permission to reproduce Figure 4.4, after figure 4.1 in R.U. Cooke and J.C. Doornkamp, *Geomorphology in Environmental Management: A New Introduction* (Clarendon Press, Oxford, 1990).

Pearson Education, Harlow, for permission to reproduce Figures 2.3 and 6.1, after figures 9.27 and 8.7 in M. Summerfield, *Global Geomorphology: An Introduction to the Study of Landforms* (Longman, Harlow, 1991).

Province of British Columbia, Victoria, for permission to reproduce Plates 8.3 and 8.4, reprinted from BC Air Photos BC207-55 and BC93026-77.

River Restoration Centre, Silsoe, for permission to reproduce in adapted form Figure 10.4(d) and Box 10.2 Figure 1 from figure 3.2.1 and 'The River Skerne Restoration Project' (p. 5) in River Restoration Centre, *River Restoration Manual of Techniques* (River Restoration Centre, Silsoe, 1999).

Mike Simms for permission to reproduce Plates 4.2 and 7.2.

South Africa National Parks and *Centre for Water in the Environment,* University of the Witwatersrand, Johannesburg, for permission to reproduce Plate 8.14.

Springer Science, Dordrecht, for permission to reproduce Figure 10.1, after figure 1.1 in G.E. Petts and C. Amoros, *Fluvial Hydrosystems* (Chapman & Hall, London, 1996).

Taylor & Francis, London, for permission to reproduce Figure 4.3, after figure 2.6 in R.J. Hugget, *Fundamentals of Geomorphology* (Routledge, London, 2003), and Box 6.2 Figure 3, after figure 3.9 in M. Kay, *Practical Hydraulics* (E. & F.N. Spon, London, 1998).

United States Geological Survey, Washington DC, for permission to reproduce Box 7.4 Figure 1, after figure 1 in W.W. Emmett, *A Field Calibration of the Sediment-trapping Characteristics of the Helley–Smith Bedload Sampler*, US Geological Survey Professional Paper 1139 (1980).

Tony Waltham Geophotos for permission to reproduce Plate 7.5 and Colour Plate 13.

Every effort has been made to contact copyright holders for their permission to reprint material in this book. The publishers would be grateful to hear from any copyright holder who is not here acknowledged and will undertake to rectify any errors or omissions in future editions of this book.

1

INTRODUCTION

The term fluvial is derived from the Latin *fluvius*, meaning river. Fluvial geomorphology is the study of the interactions between river channel forms and processes at a range of space and time scales. The influence of past events is also significant in explaining the present form of river channels. Rivers are found in many different environments and show an amazing diversity of form. In this chapter you will learn about:

- The different scales at which rivers are examined.
- How the form of a river channel is determined by the balance between the physical force acting on the material forming the channel and the resistance of that material to being moved.
- The way in which local environmental factors interact with these hydraulic forces to produce a wide variety of channel forms.

DIVERSITY OF FORM

A quick look through the photographs in this book will give you some idea of the variety that can be seen in rivers and streams worldwide. Rivers drain much of the land area – with the exception of regions that are hyper-arid or permanently frozen – and their variety reflects the vast range of different environments in which they are found. Climate, geology, vegetation cover and

topography are just some of the factors that influence river systems.

Rivers are found in many different climatic zones, ranging from humid to arid, and from equatorial to arctic. Some of the larger rivers even flow across different climatic zones, originating in a humid area before flowing through an arid region. Examples of these 'exotic' rivers include the Nile and Colorado, both of which sustain agriculture and urban centres in desert regions. **Perennial** rivers flow for all or most of the year, while many of those in dryland environments only transmit water at certain times. The 'trail' shown in Colour Plate 1 is actually an **ephemeral** channel that was photographed in South Africa during the winter dry season. A small herd of cattle in the distance provide an idea of scale.

The material in which the channel is formed is called the **channel substrate**. An important distinction can be made between bedrock and alluvial substrates (Figure 1.1). **Bedrock channels**, as their name suggests, are sections of channel that are cut directly into the underlying bedrock, while **alluvial channels** are formed in alluvium – sediment that has previously been laid down in the valley floor by rivers. Alluvium can include a mixture of unconsolidated particles ranging in size from boulders, gravels and sands to finer deposits of silts and clays. Where the valley floor is wide enough, material laid down in the channel, together with silt deposited by

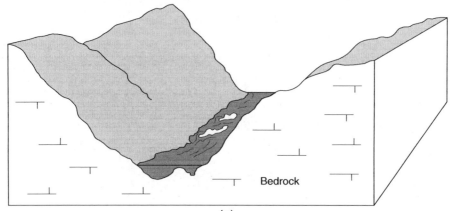

(a)

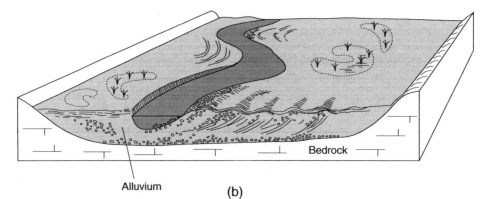

(b)

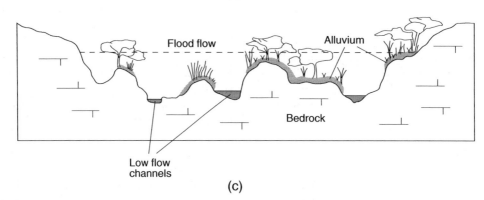

(c)

Figure 1.1 (a) Bedrock channels are cut directly into the underlying bedrock. (b) Alluvial channels are formed in alluvial deposits that have been deposited on the valley floor by fluvial processes. (c) Rivers exist on a continuum from bedrock to alluvial and there are also many mixed alluvial/bedrock channels. This is just one example, showing a cross-section of a mixed channel reach. Bars separate the flow into multiple channels. Each bar has a bedrock core, which is overlain by vegetated alluvial deposits. (c) adapted from van Niekirk *et al.* (1999).

floods, form a **floodplain** adjacent to the river channel (Figure 1.1b). The ephemeral channel shown in Colour Plate 1 is formed in alluvium. Colour Plate 2 shows an example of a bedrock channel, while Colour Plate 3 shows a mixed channel which has a rock bed and alluvial banks.

Most rivers flow to the oceans, although some drain to inland seas and lakes, while others dry up completely before reaching the ocean. Each river drains an area of land called its **drainage basin** – also known as its **catchment** or **watershed** – which supplies water and sediment to the channel (see Figure 1.2). This area is bounded by a **drainage divide** or **catchment boundary**, something that is clearly visible as a ridge in mountainous areas but which can be rather difficult to discern in areas of more subdued topography. The outlet, where the main channel exits the basin, is at a lower elevation than the rest of the basin area. Drainage basins form a mosaic across the land surface, varying greatly in size from a few hectares to millions of square kilometres.

Within each drainage basin is a branching network of channels. The main, or trunk, channel is fed by numerous small tributaries which join to form progressively larger channels. Drainage patterns, as viewed from the air or on maps, vary considerably between basins. The development and evolution of drainage networks is influenced by a number of factors, including geology, climate and long-term drainage basin history. Further information on the ways in which structural controls influence drainage patterns can be found in Box 1.1.

In terms of the actual form of different channels, obvious differences can be seen, even along the same river. One of the more noticeable things is the variation in size, from tiny headwater streams that are just a few centimetres wide to large rivers several hundred metres or more in width. The size of a river channel at a given point is largely determined by the **discharge** supplied from upstream. This is the volume of water that passes through a given channel cross-section in a given period of time. In the upper reaches of a river, the area drained – and hence the discharge – is relatively small. As you move downstream, discharge and channel size generally increase with the upstream drainage area.

While many rivers follow a single channel, there are also numerous examples of rivers with multiple channels. Those that flow in a single channel usually tend to

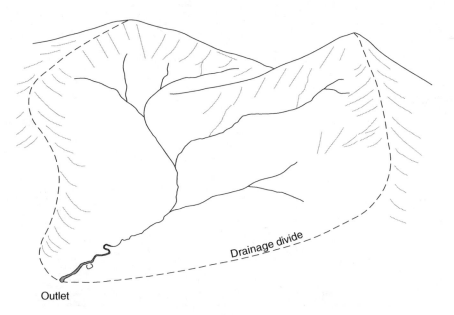

Figure 1.2 The drainage basin is the area of land drained by each river.

Box 1.1

DRAINAGE NETWORK PATTERNS

Water and sediment are transported from the hillslopes to the basin outlet via the drainage network. The shape and density (length of stream channel per unit area) of drainage networks varies considerably, and no two are the same. Figure 1 provides some typical examples of selected drainage patterns. A major control on drainage network morphology is the underlying geology, although topography, soils, tectonic history and climate are also influential. **Dendritic** drainage networks, which have a random pattern, are found where there are no strong geological controls. Where there is a strong regional dip (slope), **parallel** drainage patterns develop. **Trellis** networks are also associated with regional dip, although in this case structural controls are also important; for example, where lines of weakness are provided by well developed joints running at right angles to the dip. Where two sets of structural controls run at right angles to each other, a **rectangular** pattern develops. Trellis and rectangular networks may also be found where there are alternating bands of hard and soft strata. **Radial** drainage develops as result of symmetrical erosion around uplifted domes and volcanoes. Although drainage patterns are used in the interpretations of tectonic influences and underlying structural controls, there may not always be a close correspondence. During their long term history, rivers cut down through great thickness of rock, and former drainage patterns are sometimes preserved.

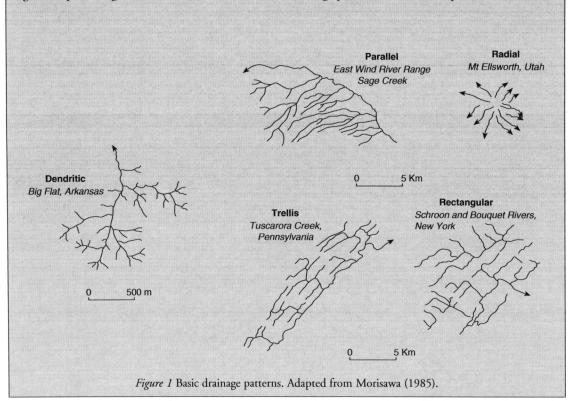

Figure 1 Basic drainage patterns. Adapted from Morisawa (1985).

deviate from a straight line, sometimes following rather an irregular path, in other cases forming more regular **meanders** (Colour Plates 4 and 5). **Braided channels** are characterised by numerous bars and islands formed by sediment deposits in the channel (Colour Plates 6 and 7). A rather different type of multiple channel are the **anabranching channels** shown in Colour Plates 8 and 9. Rather than being separated by bars, the individual channels are cut into the floodplain itself.

When considering channel form, individual sections, or **reaches**, of channel are usually considered. This is because of the downstream changes in channel size and shape that are brought about by factors such as increasing drainage area and variations in channel substrate. As a result, different channel patterns may be found along the same channel. At the reach scale – typically a few tens to hundreds of metres or more – there is a homogeneity of form.

Rivers are three-dimensional in shape and, in addition to the channel planform patterns described above, variations are also seen in cross-sectional shape and channel slope (Figure 1.3). For example, braided channels tend to be relatively wide and shallow in comparison to meandering channels, which have a narrower, deeper cross-section. Headwater streams in mountainous areas typically flow in steep channels, with frequent waterfalls, pools and rapids. This is in contrast to rivers that flow across lowland floodplains, where the channel slope is much more gentle.

Fluvial forms also exist at the sub-channel scale. These include channel bars, pools excavated by localised scour, and periodic features such as dunes and ripples that form on the bed of sandy channels. You will see later in the book how certain groupings, or assemblages, of these features are associated with different channel types.

HOW RIVERS SHAPE THEIR CHANNELS

Rivers and streams continuously shape and reform their channels through erosion of the channel boundary (bed and banks) and the reworking and deposition of sediments. For example, erosion and undermining of the banks can lead to channel widening. Scouring of the channel bed deepens the channel, while sediment deposition reduces the depth and can lead to the formation of channel bars. These are just some of the ways in which channel adjustment takes place.

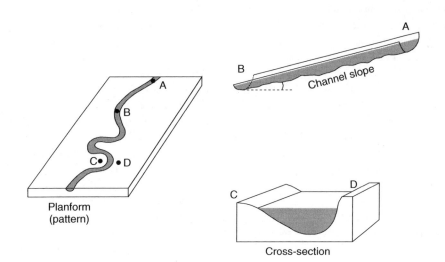

Figure 1.3 Elements of three-dimensional channel form. The planform is the shape of the river viewed from above, the channel slope is shown for the reach between points A and B, and the channel cross-section between C and D.

There is an important balance between the erosive power of the flow and the strength, or resistance, of the bed and bank material to erosion. During major flood events, when the erosive power of the flow is greatly increased, there can be dramatic changes in channel form. Just how dramatic these changes are depends on how much resistance is provided by the bed and banks. Channels formed in unconsolidated alluvium offer much less resistance to erosion than those cut in bedrock. In fact, most flows are able to shape channels formed in sandy alluvium because relatively little energy is required to set the individual sand grains in motion. Because silts and clays are smaller than sand grains, you might expect these particles to be easier to erode. However, they tend to stick together as a result of cohesive (attractive) electro-chemical forces between the particles. This means that channel boundaries with a high proportion of silt and clay are actually more resistant to erosion than those formed in sand and fine gravel. The amount and type of vegetation growing along the banks is also significant, since this can provide additional resistance to erosion.

In many cases, it is only extreme floods that are capable of significantly modifying bedrock channels. The comparative rarity of these very large floods means that channel adjustments tend to occur sporadically, being interspersed by long periods of little change. Alluvial channels dominated by cobbles and boulders may also be relatively unaffected by most flows, which are not powerful enough to move such coarse material.

The energy needed to carry out geomorphological work is provided by the flow of water through the channel. For any length of channel, energy availability is dependent on two things: the flow discharge and the steepness of the channel slope. Increases in either of these will increase the stream power and therefore the potential to carry out geomorphological work. However, before sediment transport and erosion can take place, a surprising amount of energy has to be used simply to move water through the channel. This is because of various types of flow resistance, including friction between the flowing water and the channel boundary. This can be particularly high in rough, boulder bed channels but is also significant for channels formed in finer substrates. Energy is also expended when the channel impinges against the valley walls,

when the flow moves around bends and as it cascades over steps and waterfalls. Friction is even generated within the flow itself as a result of eddies and turbulence. It is estimated that 95 per cent of a river's energy is used in overcoming flow resistance, leaving just 5 per cent to carry out geomorphological work.

Flow and sediment supply

The flow in natural channels constantly fluctuates through a continuous series of normal flows, floods and droughts. Sediment supply also varies through time. Rivers continuously adjust their form in response to these fluctuations, which in turn influences the flow of water and sediment transport through the channel.

Because the flow of water in a river provides the energy required to shape the channel, the characteristics of that flow are very important in determining channel form. As previously mentioned, the mean discharge usually increases with the size of the upstream drainage area. However, the mean discharge does not reflect the way in which flow varies through time. These variations are described by the **flow regime**, which can be thought of as the 'climate' of a river. Characteristics of the flow regime include seasonal variations in flow and the size and frequency of floods.

Processes of erosion, transport and deposition within a channel reach are influenced by the supply of sediment at the upstream end as well as sediment that is locally eroded from the bed and banks. It is not only the volume of sediment that is important, but also its size distribution. Processes of sediment transport are very different for coarse and fine sediment, so sediment supply has an important influence on channel form and behaviour. The finer materials – clay particles, silts and sands – are carried in the flow as **suspended load**. This can be transported over considerable distances. Coarse sediment, because it weighs more, is transported close to the channel bed as **bedload**. Compared with the suspended load, bedload movement is more localised, involving much shorter travel distances. Deposits of coarse material form the channel bars that characterise many alluvial and bedrock channels, although finer-grained sand and silt bars are also common. As you will see later in this book, the form and behaviour of bedload-dominated channels is rather different from suspended load dominated channels.

Valley setting

Channel processes are driven by flow and sediment supply, although the range of channel adjustments that are possible are often restricted by the valley setting. The influence of channel substrate and vegetation on bank erosion and channel migration have already been mentioned. The valley slope is also significant, affecting the steepness of the channel, which, together with discharge, determines stream power. Channels that flow over very gentle gradients can sometimes be extremely restricted in the adjustments they can make because so little energy is available. Another control on channel adjustment is the degree of **valley confinement**. While some channels are able to migrate freely across a wide floodplain, others are confined to a greater or lesser extent by the valley walls. Various degrees of valley confinement can be seen in Colour Plates 4 to 11.

THE FORM OF A CHANNEL

With so many environmental variables influencing channel form, an enormous range of different channel forms and behaviour is possible. It should be pointed out that not all rivers fit neatly into one of these categories – there are many examples of transitional rivers that have characteristics associated with more than one channel type.

Alluvial channel form

Four main types of alluvial channels are generally recognised: straight, meandering, braided and anabranching.

Straight channels

Although there are many examples of streams and rivers that have been artificially straightened for engineering purposes, naturally straight channels are rare. Even where they do exist, variations are usually seen in flow patterns and bed elevation. Straight channels are relatively static, with rates of channel migration limited by a combination of low energy availability and high bank strength. This is especially true where the channel banks are formed from more resistant material, such as cohesive silts and clays.

Meandering channels

Meanders form in a variety of bedrock and alluvial substrates. Associated with moderate stream powers, alluvial meanders may develop in gravels, sands, or fine-grained silts and clays. An interesting characteristic of meanders is that they are scaled to the size of the channel, being more widely spaced for larger channels.

The degree of meandering varies greatly, from channels that only deviate slightly from a straight line to sequences of highly convoluted meander bends. Variations are also seen in the regularity of meander bends, many of which are rather more irregular than the examples shown in Colour Plates 4 and 5. Meandering channels evolve over time as individual bends migrate across the floodplain. Erosion is usually focused at the outside of meander bends, which gradually eat into the floodplain as the channel migrates laterally. At the same time, deposition on the inside of the bend allows the channel to maintain its width. Cut-offs – short sections of abandoned channel – indicate the path of former meander bends (Colour Plate 11).

Braided channels

Braided rivers are characterised by wide, relatively shallow, channels in which the flow divides and rejoins around bars and islands (Colour Plates 6 and 7). The appearance of a braided channel varies with changing flow conditions. During high flows, many of the bars become partly or wholly submerged, giving the appearance of a single wide channel. At low flows, extensive areas of bar surface may be exposed (Colour Plate 7).

In order for bars to form, an abundant supply of bedload is required. Much of this is supplied from the upstream catchment area, with additional contributions from bank erosion. The bars themselves can be formed from sand, gravel or boulders.

Braided rivers, are associated with high rates of energy expenditure, which is involved in the transport of large volumes of sediment. They often have steep channel slopes, although there are several examples of large braided rivers that flow over low gradients, such as the lower reaches of the vast Brahmaputra River in India and Bangladesh (Colour Plate 12). Erodible banks are also required for the channel to become wide enough to allow for the growth and development of channel bars.

Braided channels are highly dynamic, with frequent shifts in channel position. Modifications, such as the dissection and reworking of bars and the formation and growth of new bars, occur over relatively short periods of time (days to years). The presence of bars leads to complex patterns of flow within the channel, and there can be sudden shifts in the location of sub-channels. Individual channels can be abandoned or reoccupied in the space of a few days.

Anabranching channels

Anabranching channels, where the flow is divided into two or more separate channels, are relatively rare in comparison to braided and meandering channels. The separate channels, called **anabranches**, are typically cut into the floodplain, dividing it up into a number of large islands. Individual anabranches can themselves be straight, meandering or braided.

Unlike braided channels, rates of lateral channel migration are typically very low. The islands are stable features and, depending on climatic conditions, are often well vegetated. However, new channels can be cut when floodwaters breach the channel boundary and spill out on to the floodplain. Other channels are abandoned as the flow is diverted elsewhere, or when they become infilled with sediment.

Colour Plates 8 and 9 both show examples of a subset of low-energy anabranching channel that is referred to as **anastomosing**. Although most research on anabranching channels has so far been focused on anastomosing channels, anabranching channels represent the most diverse of the four main channel types.

Bedrock channels

Bedrock channels also show a wide diversity of form. In comparison with alluvial channels, bedrock and mixed bedrock–alluvial rivers have received relatively little attention until recently. These channels often behave in a different way to alluvial channels, being strongly influenced by the resistant nature of their substrate. Structural controls, such as joints, bedding planes and the underlying geological strata can all have a significant effect on flow processes and river morphology.

As with alluvial channels, the flow may follow single or multiple channels. Straight reaches are often associated with structural controls, for example where the channel follows the line of a fault or joint. However, flow characteristics also have an influence in shaping the channel. Colour Plate 13 shows the regularly undulating walls of a slot canyon, which have been shaped by flash floods. Meanders can also form in bedrock-influenced channels, as can be seen from the spectacular incised meanders of the Colorado (Colour Plate 5). Because of the resistant substrate, bedrock meanders tend to be scaled to higher flows than their alluvial counterparts.

An example of a multi-channel bedrock river is shown in Colour Plate 2. The individual channels have cut their course to flow around bedrock bars. In some mixed bedrock–alluvial channels, bedrock bars may form a core that becomes covered in alluvial deposits, giving the appearance of an alluvial channel.

Chapter summary

Rivers are found in many different climate zones. They cut their channels into a range of bedrock and alluvial substrates. Each river drains an area called its drainage basin and is fed by a network of channels, which transports water and sediment from the land area to the basin outlet. The form of a given reach (length) of channel is controlled by the supply of flow and sediment to its upstream end. Also significant are the channel substrate, valley width, valley slope and bankside vegetation. All these controls vary, both between rivers and along the same river. This creates a huge range of fluvial environments and resultant channel forms. The three-dimensional shape of a river is described in terms of its planform, slope and cross-sectional shape. Rivers continuously adjust their channels in response to fluctuations in flow and sediment supply. An important balance exists between the erosive force of the flow and the resistance of the channel boundary to erosion. Four main types of alluvial channel form can be identified: straight, meandering, braided and anabranching. Bedrock channels also exhibit a wide variety of different forms.

FURTHER READING

Introductory texts

Many introductory geomorphology text books contain good chapters on fluvial geomorphology. These will provide a general overview and may be helpful later on if you would like a more basic explanation of certain concepts.

Gilvear, D.J., 2005. Fluvial geomorphology and river management. In: J. Holden (ed.), *An Introduction to Physical Geography and the Environment*. Pearson Education, Harlow, pp. 327–55. Well written and easy to follow, this clearly explains the key concepts and has an interesting management perspective.

Huggett, R.J., 2003. *Fundamentals of Geomorphology*. Routledge, London. The chapter on fluvial landscapes provides a good, accessible introduction.

Summerfield, M.A., 1990. *Global Geomorphology: An Introduction to the Study of Landforms*. Longman, Harlow, An excellent book, which contains clearly explained chapters on fluvial processes and fluvial landforms.

Textbooks for further reference

There are a number of textbooks of fluvial geomorphology. Most of them are written for senior undergraduate and postgraduate students.

Knighton, D.A., 1998. *Fluvial Forms and Processes: A New Perspective*. Arnold, London. A well-respected book, written at a more advanced level, which will provide valuable further reading.

Leopold, L.B., 1994. *A View of the River*. Harvard University Press, Cambridge MA. An excellent book, written for a wide audience and liked by my own students.

Leopold, L.B., Wolman, M.G. and Miller, J.R., 1964. *Fluvial Processes in Geomorphology*. Freeman, San Francisco. The 'Bible' of fluvial geomorphology. Although long out of print and available only in libraries, this was a ground-breaking book and is interesting to look through.

Morisawa, M., 1985. *Rivers*. Longman, Harlow. Also out of print, but provides a clear, well illustrated introduction with some interesting examples.

Petts, G.E. and Foster, I.D.L., 1985. *Rivers and Landscape*. Arnold, London. Out of print, but has clear explanations. Well worth referring to if you can get hold of a copy.

Richards, K., 1982. *Rivers: Form and Process in Alluvial Channels*. Methuen, London. Another classic out of print book. Some students may be put off by the mathematical approach but the explanations are clear. Good on sediment transport processes and channel adjustment.

Robert, A., 2003. *River Processes: An Introduction to Fluvial Dynamics*. Arnold, London. Written to complement the book by Knighton, with a focus on channel processes.

Thorne, C.R., Hey, R.D. and Newson, M.D. (eds), 1997. *Applied Fluvial Geomorphology for River Engineering and Management*. John Wiley & Sons, Chichester. An edited volume, with contributions from leading experts in the field, which aims to provide river engineers and managers with an overview of fluvial geomorphology.

2

THE FLUVIAL SYSTEM

A system is a collection of related objects and the processes that link those objects together. Within fluvial systems, objects such as hillslopes, the channel network and floodplains are linked together by the processes that move water and sediment between them. In common with other systems, the fluvial system is hierarchical, in that there are integrated sub-systems operating within it. This chapter will provide an overview of the fluvial system. Simplified examples are used here because the process–form interactions that are discussed in later chapters have not yet been covered. In this chapter you will learn about:

- How energy and materials move through the system.
- Fluvial system variables (e.g. channel slope, discharge, bedload transport rate).
- The way in which some variables control others and how these relationships depend on the space and time scales considered.
- How feedbacks between variables can counteract or enhance system response to change.
- The role of thresholds in system behaviour.
- How equilibrium can be defined at different time scales.

INPUTS, OUTPUTS AND STORES

The basic unit of the fluvial system is the drainage basin. Fluvial systems are open systems, which means that energy and materials are exchanged with the surrounding environment. In closed systems, only energy is exchanged with the surrounding environment.

Inputs

The main inputs to the system are water and sediment derived from the breakdown of the underlying rocks. Additional inputs include biological material and solutes derived from atmospheric inputs, rock weathering and the breakdown of organic material.

Most of the energy required to drive the system is provided by the atmospheric processes that lift and condense the water that falls as precipitation over the drainage basin. The pull of gravity then moves this water downslope, creating a flow of energy through the system. This energy is expended in moving water and sediment to river channels and through the channel network.

Outputs

Water and sediment move through the system to the drainage basin outlet, where material is discharged to the ocean. Not all rivers reach the ocean; some flow into inland lakes and seas, while others, such as the Okavango River in Botswana, dry up before reaching the ocean. This reflects another important output from fluvial systems: the loss of water by evaporation to the atmosphere.

Most of the available energy is used in overcoming the considerable frictional forces involved in moving water and sediment from hillslopes into channels and through the channel network. Much of this energy is 'lost' to the atmosphere in the form of heat.

Stores

A certain amount of material is stored along the way. For example, water is stored for varying lengths of time in lakes and reservoirs, and below the ground in the soil and aquifers. Sediment is stored when it is deposited in channels, lake basins, deltas, alluvial fans and on floodplains. This material may be released from storage at a later stage, perhaps when a channel migrates across its floodplain, eroding into formerly deposited sediments which are then carried downstream. Ferguson (1981) describes the channel as 'a jerky conveyor belt', since sediment is transferred intermittently seawards.

TYPES OF SYSTEM

Three types of system can be identified in fluvial geomorphology. These are morphological systems, cascading systems and process–response systems.

Morphological (form) systems

Landforms such as channels, hillslopes and floodplains form a morphological system, also referred to as a **form system**. The form of each component of a morphological system is related to the form of the other components in the system. For example, if the streams in the headwaters of a drainage basin are closely spaced, the hillslopes dividing them are steeper than they would be if the streams were further apart from each other. Relationships such as this can be quantified statistically.

Cascading (process) systems

The components of the morphological system are linked by a cascading system, which refers to the flow of water and sediment through the morphological system. Cascading systems are also called **process systems** or **flow systems**. These flows follow interconnected pathways from hillslopes to channels and through the channel network.

Process–response systems

The two systems interact as a process–response system. This describes the adjustments between the processes of the cascading system and the forms of the morphological system. There is a two-way feedback between process and form. In other words, processes shape forms and forms influence the way in which processes operate (rates and intensity). This can be seen where a steep section of channel causes high flow velocities and increased rates of erosion. Over time erosion is focused at this steep section and the channel slope is reduced. Velocity decreases as a result, reducing rates of erosion.

In order to examine the components of the fluvial system in more detail, it can be divided into sub-systems, each operating as a system within the integrated whole. One way of doing this is to consider the system in terms of three zones, each of which is a process–response system with its own inputs and outputs (Figure 2.1). Within each zone certain processes dominate. The sediment **production zone** in the headwater regions is where most of the sediment originates, being supplied to the channel network from the bordering hillslopes by processes of erosion and the mass movement of weathered rock material. This sediment is then moved through the channel network in the sediment **transfer zone**, where the links between the channel and bordering hillslopes, and hence sediment production, are not so strong. As the river approaches the ocean, its gradient declines and the energy available for sediment transport is greatly reduced in the sediment **deposition zone**. It is primarily the finest sediment that reaches the ocean, as coarser sediment tends to be deposited further upstream. In fact, only a certain proportion of all the sediment that is produced within a drainage basin actually reaches the basin outlet.

FLUVIAL SYSTEM VARIABLES

Variables are quantities whose values change through time. They include such things as drainage density, hillslope angle, soil type, flow discharge, sediment yield, channel pattern and channel depth.

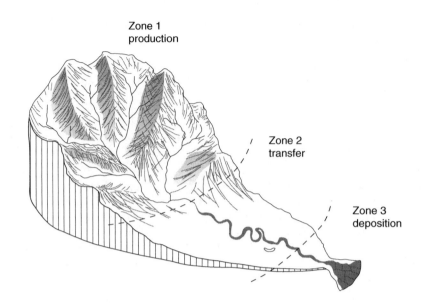

Figure 2.1 The fluvial system can be subdivided into three zones on the basis of the dominant processes operating within each zone. These are the sediment production zone, sediment transfer zone and sediment deposition zone. Adapted from Schumm (1977).

Internal and external variables

An important distinction exists between internal and external variables. All the examples given above are **internal variables**, which operate within the fluvial system. Internal variables are influenced by other internal variables, and also by variables that originate from outside the system. These **external variables**, such as climate, control or regulate the way in which the system operates. Unlike the internal variables, external variables operate independently, in that they are not influenced by what is going on inside the fluvial system. At the basin scale the external variables are climate, base level, tectonics and human activity.

If you are considering a sub-system, such as a reach of channel in the transfer zone, the external variables would include the supply of flow and sediment to the channel. This is because these variables originate from outside the channel sub-system, even though they are internal variables at the basin scale. To avoid confusion, the 'ultimate' external variables – climate, base level, tectonics and human activity – will be referred to as **external basin controls**.

The external basin controls

The variables defined in this section act as regulators of the whole system. Any change in one of these variables will lead to a complex sequence of changes and adjustments within the fluvial system.

- *Climate* describes the fluctuations in average weather. Although the weather is always changing, longer-term characteristics such as seasonal and inter-annual variations can be defined. Other characteristics include how often storms of a given size can be expected to occur and the frequency and duration of droughts. Where no long-term changes are occurring in the climate, the combination of such attributes defines an envelope of 'normal' behaviour. **Climate change** occurs when this envelope shifts and a new range of climatic conditions arises.
- *Tectonics* refers to the internal forces that deform the Earth's crust. These forces can lead to large scale uplift, localised subsidence, warping, tilting, fracturing and faulting. Where uplift has occurred,

inputs of water have to be lifted to a greater elevation, increasing energy availability; some of the highest rates of sediment production in the world are associated with areas of tectonic uplift. Valley gradients are altered by faulting and localised uplift, which may in turn affect channel pattern. Lateral (sideways) tilting can cause channel migration and affect patterns of valley sedimentation.

- **Base level** is the level below which a channel cannot erode. In most cases this is sea level. If there is a fall in sea level relative to the land surface, more energy is available to drive flow and sediment movement. Conversely a relative rise in base level means that less energy is available, resulting in net deposition in the lower reaches of the channel. Over time these effects may be propagated upstream through a complex sequence of internal adjustments and feedbacks.

- **Human activity** has had an increasing influence on fluvial systems over the last 5,000 years, especially during recent times. Activities within the drainage basin such as deforestation, agriculture and mining operations all affect the flow of water and production of sediment. These are referred to as indirect or diffuse activities. River channels are also modified directly when channel engineering is carried out. Advances in technology over the last century have meant that dam construction, channel enlargement for navigation and flood control, channel realignment, the building of flood embankments and other engineering works can now be carried out at an unprecedented scale. Today there are very few rivers that have not been affected in some way by the direct and indirect effects of human activity. It can be argued that, under some circumstances, human activity can be considered to be *both* an internal and an external variable. Many of the direct modifications described above are in response to some local human perception of the system. For example, channels are dredged because they are not deep enough for navigation, or flood defence works carried out because floods occur too frequently. Urban (2002) suggests that direct human intervention can often be classified as an internal variable, although it is more appropriate to consider *indirect* human activities as external.

The influence of the external basin controls on the fluvial system is represented schematically in Figure 2.2. Some internal variables have a greater degree of independence in that they are only affected in a limited way by the fluvial system. These variables are geology, soils and vegetation and topography (which includes relief, altitude and drainage basin size). All are internal variables because they are controlled to some extent by the external basin controls, however their main influence on the operation of the fluvial system is a controlling one.

Adjustable (dependent) and controlling (independent) variables

From the discussion above it can be seen that some variables control the adjustment of other variables. For example channel pattern is, among other things, affected by the supply of sediment to the channel. In this case, channel pattern is the **adjustable** or **dependent variable** while sediment supply is the **controlling** or **independent variable**.

Things can get a little confusing because controlling variables may in turn be adjusted by other variables. Extending the previous example, sediment supply is itself controlled by hillslope vegetation cover. In this case, sediment supply is the adjustable variable and vegetation cover the controlling variable. All internal variables are adjustable because their operation is ultimately regulated by the external basin controls. They are also influenced to a greater or lesser extent by other internal variables. Because the relationships between variables are so complicated, it can be very difficult to isolate the effect of one variable on another.

The hierarchical nature of the fluvial system means that variables operating at larger scales tend to affect the operation of variables at smaller scales. For example, climate affects vegetation cover and hillslope erosion, which in turn determine sediment supply, which influences channel pattern, which affects the small-scale flow dynamics in the channel, which governs the movement of individual grains. This is not a one-way process, however. Over long periods of time, the cumulative effect of small-scale processes, such as the erosion and deposition of individual grains, can lead to larger scale changes. These include changes in channel pattern and, over time

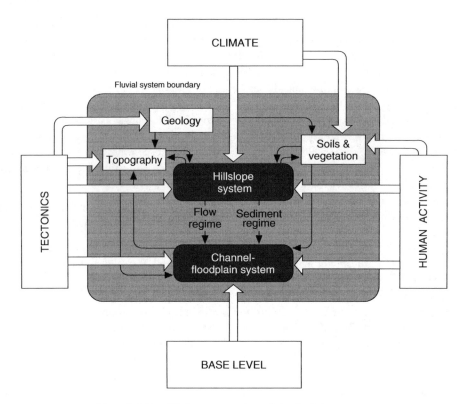

Figure 2.2 Simplified representation of the fluvial system.

periods of tens to hundreds of thousands of years, can adjust the slope of the whole river valley.

Time itself is an important controlling variable. Every drainage basin has a historical legacy resulting from past changes that have taken place in the basin. This includes the cumulative effect of processes such as erosion, transport and deposition over long periods of time. It also includes the far-reaching effects of changes in the external basin controls, such as the variations in climatic conditions since the Last Glacial Maximum 18,000 years ago, which have greatly affected fluvial systems worldwide. In the temperate zone, many rivers underwent a transition from a braided to a meandering form as climate conditions ameliorated, vegetation became established and sediment loads decreased. However, vast quantities of sediment still remain in formerly glaciated drainage basins, where many fluvial systems are still adjusting to this glacial legacy.

Feedbacks

A **feedback** occurs when a change in one variable leads to a change in one or more other variables, which acts to either counteract or reinforce the effects of the original change. Two types of feedback are observed: negative feedback and positive feedback. Both are initiated by a change in one of the system variables, which in turn leads to a sequence of adjustments that eventually counteract the effect of the original change (**negative feedback**) or enhance it (**positive feedback**).

When there is a change in one of the external controls, negative feedbacks allow the system to recover, damping out the effect of the change. An everyday example of a negative feedback loop is a central heating system controlled by a thermostat, which switches the source of heat on and off as the room cools and warms. An equilibrium is maintained as the temperature fluctuates around an average value. A commonly cited example of negative feedback within the fluvial system occurs when a section of channel is suddenly steepened by tectonic faulting. This leads to a local increase in the flow velocity and rate of bed erosion. Over time this acts to reduce the channel slope, counteracting the effects of the original change. It should be noted that the actual sequence of events is usually rather more complex. This is because change in one part of the system can lead to complex changes, both locally and throughout the rest of the system. The nature of complex response will be discussed later in this chapter.

Positive feedbacks have a very different effect. Soil erosion is a natural process and an equilibrium exists if rates of soil removal over a given period are balanced by rates of soil formation over that period. However, an external change, such as the deforestation of steep slopes, can lead to a dramatic increase in soil erosion. The upper soil layers contain the most organic matter, which is important in binding the soil together. It also increases soil permeability, allowing rainfall to soak into the soil rather than running over and eroding the soil surface. If the topsoil is removed, the lower permeability of the underlying soil layers means that more water runs over the surface, increasing erosion and removing still more soil layers. In this way several centimetres of soil can be removed by a single rain storm (Woodward and Foster, 1997). This greatly exceeds the rate of soil formation. Referred to colloquially as 'vicious circles' or 'the snowball effect', positive feedbacks involve a move away from an equilibrium state. They usually involve the crossing of a **threshold** (see below) as the system moves towards a new equilibrium. A small-scale example of positive feedback is the build up of sediment during the formation of a channel bar. Bar formation is initiated when bedload sediment is deposited at a particular location on the channel bed (the various mechanisms by which this occurs will be discussed in later chapters). This affects local flow dynamics, causing the flow to diverge over and around the initial deposit. As the flow diverges, it becomes less concentrated and therefore less able to transport the coarser sediment. Localised deposition occurs, further disrupting the flow and promoting further deposition and bar growth.

Several feedbacks, both positive and negative, exist between channel form, water flow and sediment transport.

These are quite complex and, in order to understand them more fully, you will need to be familiar with concepts introduced in Chapters 6 and 7. For this reason, a basic overview is provided here, with a more detailed description in Chapter 8.

The form of a channel has an important influence on the way that water and sediment move through it. For example, flow is concentrated where the channel narrows, increasing erosion potential. As you saw above, deposition may occur where the flow diverges around obstacles such as bars. The character of the channel bed is also significant, since the size and arrangement of sediment determines bed roughness and resistance to flow. Where resistance is high, the average velocity of flow in the channel is reduced. This influences hydraulic conditions near the bed of the channel, which are significant for processes of erosion and deposition. Considerable differences are seen across the channel bed, giving rise to spatial variations in erosion and deposition. These processes themselves modify the form of the channel, feeding back to influence flow.

Thresholds

Thresholds are another important concept in systems theory and you will come across many examples in fluvial geomorphology. For example, a threshold is crossed when a sand grain on the bed of the channel is entrained (set in motion). Movement is resisted by the submerged weight of the grain and friction between it and the neighbouring grains. If the driving force exerted on the grain by the flow is less than these resisting forces, no movement will occur. It is only when the driving force of the flow exceeds the submerged weight of the grain that entrainment will take place. In this example, channel flow is an external variable.

When a threshold is crossed there is a sudden change in the system, for example when loose material on a slope becomes unstable and starts to move down the slope as a landslide. The gradual processes by which rock is broken down and loose material builds up on a hillslope take place over time scales of tens to hundreds of years. Why, then, does a landslide occur at a particular point in time? Such a transition can come about when a change in one of the controlling variables leads to instability within the system – as a direct result of an earthquake (tectonics) for example. Thresholds that are crossed as a result of external change are called **external thresholds**. Instabilities may also develop over time without any external change having occurred. For this reason it is possible for a major landslide to be triggered by a relatively minor rainfall event that falls well within the expected climatic norm, because instability has gradually developed over time and the system is ready 'primed'. This is an example of an **internal threshold**. Another commonly cited example is the threshold that is crossed when a meander loop is cut off to form an oxbow lake (several of these are shown in Colour Plate 11). Again this is something that can take place without there having been any change in the external variables.

Whether or not either kind of threshold is crossed depends on how 'sensitive' the system is, in other words how close to a threshold it is. To illustrate this point, consider a pan filled with water that is heated by 10°C at normal atmospheric pressure. If the water had an initial temperature of 25°C you would not expect to see much change in its appearance. However, if the initial temperature was 90°C, the same increase in temperature would lead to a dramatic change as a threshold was crossed and the water started to boil. In the second example, the sensitivity of the system is much greater because the water temperature is closer to boiling point. Once a threshold has been crossed, the system reaches a new equilibrium. Relating this back to the sand grain example, two scenarios could be considered. In the first, the particle is resting on a flat bed and is fully exposed to the flow. In the second, it is buried beneath a layer of gravel. The threshold for movement will be much lower for the exposed grain than for the buried grain, which cannot move unless the overlying gravel is removed.

Complex response

The response of the fluvial system to change is often complex because of the many interrelationships that exist between the different components of the system. An example is the complex response of a tributary to a lowering in base level elevation at its outlet (Schumm, 1977). Here the main river, into which the tributary flows, has degraded, or lowered its channel elevation by erosion.

This leads to a complex sequence of episodic erosion and deposition in the tributary as the system searches for a new equilibrium. A similar sequence of events can be observed from experimental simulations of laboratory drainage networks (Summerfield, 1991) and is illustrated schematically in Figure 2.3. This represents a sequence of events that occurs at the upstream end of a channel (boxes running down the right hand side of the figure) and at the downstream end (boxes running

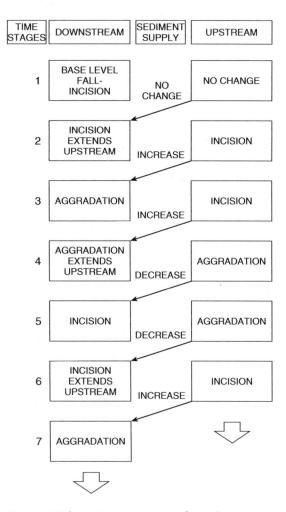

Figure 2.3 Schematic representation of complex response in a drainage system as a result of a fall in base level. See text for full explanation. After Summerfield (1991).

down the left hand side of the figure). The numbers on the left hand side (1 to 7) indicate sequential time stages. Starting at the top of the page and working down, we can see what is happening upstream and downstream at each time stage by reading across the page. Changes in sediment supply from upstream are also indicated by the arrows running from 'upstream' boxes to 'downstream' boxes.

- *Stage 1.* A fall in base level results in a local steepening of the channel gradient at the downstream end. This might exist as a sharp break in channel slope or as a steeper section of channel. Erosion is focussed at this steeper part, leading to a localised lowering in channel elevation. As a result of this lowering, the section of channel immediately upstream becomes steeper. The steep zone therefore migrates upstream over time. The initial response of the channel to the fall in base level is therefore incision (downcutting) at the downstream end. At this stage the upstream part of the channel is unaffected, so there is no change in the supply of sediment coming from upstream.
- *Stage 2.* Over time, incision is propagated upstream. As a result there is an increase in the amount of sediment coming from further upstream.
- *Stage 3.* This increase in sediment supply from upstream leads in turn to a build up of sediment in the channel at the downstream end. This process is called **aggradation** and results in a reduction in channel gradient. Meanwhile, the zone of incision continues to migrate further upstream.
- *Stage 4.* Aggradation is propagated upstream. Since there is now a net deposition of sediment in the upstream part of the channel, there is a decrease in the volume of sediment supplied from upstream.
- *Stage 5.* At the downstream end, less sediment means that there is additional energy for erosion and a new cycle of incision begins. Aggradation continues to occur at the upstream end.
- *Stage 6.* The wave of incision extends upstream, once again increasing the supply of sediment from upstream.
- *Stage 7.* The increased sediment supply leads to aggradation at the downstream end of the channel.

This example illustrates the way in which downstream changes can be propagated upstream, while the resultant upstream changes in turn control what happens downstream. In this way there is a complex cycle as periods of **episodic erosion** are interspersed with periods of deposition.

SCALE IN FLUVIAL GEOMORPHOLOGY

Scale is an important consideration in fluvial geomorphology, with process–form interactions occurring over a huge range of space and time scales. At one end of this range is the long-term evolution of the landscape. At the other are small-scale processes, such as the setting in motion of an individual grain of sand resting on the bed of a channel. Space scales therefore encompass anything from a few millimetres to hundreds of kilometres. Relevant time scales stretch from a few seconds to hundreds of thousands of years or more.

In order to understand how the fluvial system operates we can examine the relationships between processes and form in more detail at finer scales. This can be done by examining individual sub-systems, or sub-systems within sub-systems. When focusing in like this it is important to remember that these sub-systems are all part of an integrated whole and therefore cannot be considered in isolation from the rest of the system.

Space scales (spatial scales)

In studying the fluvial system, the scale of relevance varies according to the type of investigation. At the largest, drainage basin scale (Figure 2.4a), it is possible to see the form and characteristics of the drainage network and drainage basin topography. These reflect the cumulative effect of processes operating over long time scales, as well as past changes imposed by the external basin controls.

At a smaller scale, the form of a reach of meandering channel can be examined in the context of drainage basin history and the influence of controlling variables at the channel scale, such as the supply of water and sediment from upstream (Figure 2.4b). The way in which the form and position of the channel has changed over time scales extending to thousands of years may be preserved as floodplain deposits, which can be used in reconstructing drainage basin history.

Moving in to look at an individual meander bend (Figure 2.4c), process–form interactions can be observed at a smaller scale. These include flow hydraulics within the bend and associated sediment dynamics. Investigations of rates of bend migration or bank erosion processes are also carried out at this scale.

Depositional channel units such as the point bar (seen on the inside of the meander bend in Figure 2.4c) are of interest to sedimentologists, providing evidence about the flows that formed them. At a finer scale still are individual ripples on the bar surface (Figure 2.4d) formed by

the most recent high flow and, moving even closer, the internal arrangement of grains (Figure 2.4e). At the finest scale are individual grains of sediment (Figure 2.4f).

Time scales (temporal scales) and equilibrium

At smaller spatial scales, process–form interactions generally result in more rapid adjustments. At the largest scale, the long term evolution of channel networks occurs over time scales of hundreds of thousands of years or more, while the migration of individual meander bends can be observed over periods of years or decades, and small-scale flow-sediment interactions within minutes.

The perspective of the historically oriented geomorphologist concerned with the large-scale, long-term evolution of landforms is therefore very different to that of the process geomorphologist or engineer who is interested in the operation of channel processes at much shorter time scales (Schumm, 1988). Historical studies show that the fluvial system follows an evolutionary sequence of development that is interrupted by major changes induced by the external basin controls. However, over the much shorter time periods involved in the field measurement of processes, there may be little or no significant change in fluvial landforms. This might not matter too much if flow–sediment interactions at very small scales are of interest, although basin history certainly does have an influence at the reach scale, since channel form has been shaped by past changes in flow and sediment supply.

The precise definition of equilibrium is also time dependent. Equilibrium refers to a state of balance within a system, or sub-system. Negative feedback mechanisms help to maintain the system in an equilibrium state, buffering the effect of changes in the external variables. However, different types of equilibrium may exist at different time scales. These were defined by Schumm (1977) with reference to changes in the elevation of the bed of a river channel above sea level. If you were to observe a short section of river channel over a period of a few hours you would not see any change in its form (unless there happened to be a flood), although you might see some sediment transport. Over this short time period the channel is said to be in a state of **static equilibrium** (Figure 2.5a).

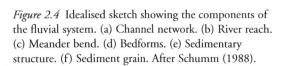

Figure 2.4 Idealised sketch showing the components of the fluvial system. (a) Channel network. (b) River reach. (c) Meander bend. (d) Bedforms. (e) Sedimentary structure. (f) Sediment grain. After Schumm (1988).

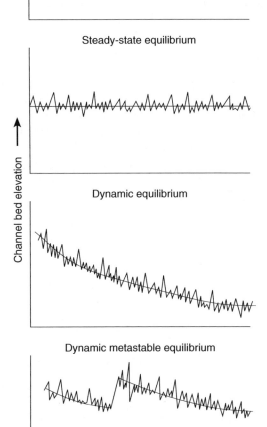

Static equilibrium

Steady-state equilibrium

Dynamic equilibrium

Dynamic metastable equilibrium

Channel bed elevation →

Time →

Figure 2.5 Types of equilibrium observed over different time scales. Adapted from Schumm (1977).

The same river, observed over a longer time scale of a decade, would show some changes. During this time, floods of various sizes pass through the channel, scouring the bed. In the intervening periods, deposition builds up the channel bed again. As a result of these cycles of scour and fill, the elevation of the channel bed fluctuates around a constant average value (Figure 2.5b) and **steady state equilibrium** exists.

Over longer time scales, from thousands to hundreds of thousands of years or more, erosion gradually lowers the landscape. At these time scales, the channel elevation fluctuates around a changing average condition, the underlying trend being a reduction in channel elevation. This is called **dynamic equilibrium** and is illustrated in Figure 2.5c.

As you know, the influence of the external basin controls cannot be ignored. Changes in any of these variables can lead to positive feedbacks within the system and a shift to a new equilibrium state. For example, in tectonically active regions, the section of channel might be elevated by localised uplift. Such episodes of change occur over much shorter time scales than the gradual evolution of the landscape, resulting in abrupt transitions. This type of equilibrium delights in the term **dynamic metastable equilibrium** (Figure 2.5d).

CHAPTER SUMMARY

A system is a group of related objects that are linked by processes. The basic unit of the fluvial system is the drainage basin. Three systems exist within this: morphological, cascading and process–response systems. Process–response systems describe the interactions between the processes (flows of energy and materials) of the cascading system and the forms of the morphological system. Fluvial system variables include hillslope angle, flow discharge and sediment transport rates. These are all quantities that change through time. Some variables (called controlling variables) act to control others (called adjustable variables). The ultimate controls are the external basin controls (climate, tectonics, base level and human activity). These influence the entire fluvial system. Complex interactions occur between internal variables as a result of feedbacks. Negative feedbacks act to mitigate the effects of a disturbance to the system, allowing a state of equilibrium to be maintained. Positive feedbacks lead to instability, exacerbating the effects of disturbances. This involves the crossing of one or more thresholds, causing the system to move towards a new equilibrium.

FURTHER READING

Introductory texts

Huggett, R.J., 2003. *Fundamentals of Geomorphology*. Routledge, London. The first chapter includes an easy to follow discussion of systems in geomorphology.

Summerfield, M.A., 1990. *Global Geomorphology: An Introduction to the Study of Landforms*. Longman, Harlow. See the first chapter for a good introduction to systems in geomorphology.

Schumm, S.A., 1977. *The Fluvial System*. John Wiley & Sons, New York. Although out of print, this provides clear explanations of the main concepts.

Schumm, S.A., 1988. Variability of the fluvial system in space and time. In: T. Rosswall, R.G. Woodmansee and P.G. Risser (eds), *Scales and Global Change: Spatial and Temporal Variability in Biosheric and Geospheric Processes*. SCOPE: 35. John Wiley & Sons, Chichester, pp. 225–50. Highly recommended, this provides a good overview of the fluvial system. Also available on line at: http://www.icsu-scope.org/downloadpubs/scope35/chapter12.html.

Classics

Schumm, S.A. and Lichty, R.W., 1965. Time, space and causality in geomorphology. *American Journal of Science*, 263: 110–19. Examines how the significance of different variables and their interrelationships varies according to the time scale under consideration.

3

THE FLOW REGIME

The flow in river channels exerts hydraulic forces on the boundary (bed and banks). An important balance exists between the erosive force of the flow (driving force) and the resistance of the boundary to erosion (resisting force). This determines the ability of a river to adjust and modify the morphology of its channel. One of the main factors influencing the erosive power of a given flow is its discharge: the volume of flow passing through a given cross-section in a given time. Discharge varies both spatially and temporally in natural river channels, changing in a downstream direction and fluctuating over time in response to inputs of precipitation. Characteristics of the flow regime of a river include seasonal variations in discharge, the size and frequency of floods and frequency and duration of droughts. The characteristics of the flow regime are determined not only by the climate but also by the physical and land use characteristics of the drainage basin. In this chapter you will learn about:

- The pathways taken by water as it travels to the channel network.
- How the flow in a river responds to inputs of precipitation.
- Seasonal variations in flow that characterise different climatic zones.
- The size (magnitude) and frequency characteristics of floods.
- Flows that are significant in shaping the channel.

FLOW GENERATION

Hydrological pathways

Inputs of water to a drainage basin are the various forms of **precipitation** that fall over its area. These include rain, snow, sleet, hail and dew. If you look at a drainage basin on a map, you will see that river channels cover only a very small part of the total area. This means that most of the water reaching the ground surface must find its way from the hillslopes and into the channel network. A number of pathways are involved (Figure 3.1), and a given 'parcel' of water arriving at a stream channel may have taken any number of them. Not all the incoming precipitation actually makes it to the basin outlet, since a certain percentage is evaporated back to the atmosphere. Water that falls on the leaves of vegetation and artificial structures like buildings is **intercepted**. Some of this water does fall to the ground, as anyone who has sheltered under a tree will know, but much of it is evaporated (water resource managers refer to this output as 'interception loss'). The amount of interception that occurs is dependent on such factors as the extent and type of vegetation (leaf size, structure, density and the arrangement of foliage), wind speed and rainfall intensity (Jones, 1997). Evaporation also takes place from the surface layers of the soil, and from lakes and wetland areas. A related process is **transpiration**,

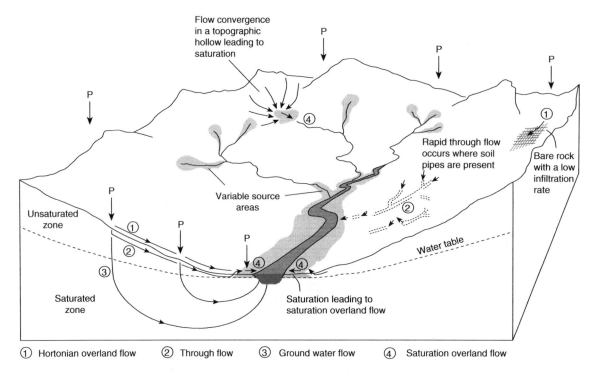

Flow convergence
in a topographic
hollow leading to
saturation

Rapid through flow
occurs where soil
pipes are present

Bare rock
with a low
infiltration
rate

Variable source
areas

Unsaturated
zone

Water table

Saturated
zone

Saturation leading to
saturation overland flow

① Hortonian overland flow   ② Through flow   ③ Ground water flow   ④ Saturation overland flow

Figure 3.1 Surface and subsurface hydrological pathways.

whereby plants take up water through their roots and evaporate it through pores, called stomata, on the underside of their leaves. This means that water can be 'lost' to the atmosphere from some depth below the surface. Because evaporation and transpiration are difficult to monitor separately, they are usually considered together as the combined process of **evapotranspiration**.

Water reaching the soil surface may either enter the soil by a process called **infiltration** or remain at the surface, moving down-slope as **overland flow** (labelled 1 and 4 in Figure 3.1). Infiltrated water either travels through the soil, parallel to the surface, as **throughflow** (2), or slowly percolates downwards to the saturated zone, travelling as **groundwater flow** (3). Rates of water movement are fastest for overland flow (typically between 50 m and 500 m per hour), between 0.005 m and 0.3 m per hour for throughflow and 0.005 m to 1.5 m per *day* for groundwater flow (Ward and Robinson, 1990). How fast a river rises after

rainfall is very much dependent on the relative proportion of water taking faster (surface and near-surface) pathways, and slower (subsurface) pathways. An important control on this is the rate at which water infiltrates into the soil, which is determined by the **infiltration capacity**. This is defined as the maximum rate at which water enters the soil when it is in a given condition (Horton, 1933). This is an important definition: if the rainfall intensity exceeds the infiltration capacity, the soil cannot absorb all the water and there is excess water, which ponds at the surface. This moves down slope as overland flow, more specifically, **Hortonian overland flow** (1). As Horton's definition implies, soil properties control the infiltration capacity; these include soil permeability, the presence of vegetation and plant roots and how much water is already in the soil. Following a dry period, the infiltration capacity is highest at the start of a storm, rapidly decreasing as rainfall continues, until a constant infiltration rate is reached (Horton, 1933).

A second type of overland flow, **saturation overland flow** (4), is generated when the soil is totally saturated and therefore cannot take any more water (Kirkby and Chorley, 1967). Where the water table is relatively shallow, for example in valley bottoms, rainfall can cause it to rise to the ground surface. Where this occurs, a saturated area forms around the channel, increasing in extent as the storm progresses. The saturated area acts as an extension to the channel network, meaning that a significant volume of water is transferred in a short time. These saturated areas are known as **variable source areas** or **dynamic contributing areas** and are highly significant in humid environments, where this is the main way in which storm run-off is generated (Hewlett and Hibbert, 1967; Dunne and Black, 1970). Hortonian overland flow is rarely observed in humid environments, unless the surface has a low infiltration rate, for example where there are outcrops of bare rock or artificially paved surfaces. However, in dryland environments, a combination of factors mean that Hortonian overland flow is the dominant mechanism. Rainfall, when it occurs, typically has a high intensity and exceeds the low infiltration capacity of sparsely vegetated soils (Dunne, 1978). In addition, dryland soils often develop a crust at the surface, which further reduces the infiltration capacity.

Water that infiltrates the soil may travel at various depths below the surface. While it is fairly obvious that water should move downwards under the force of gravity, it might seem counter-intuitive that it also flows through the soil as throughflow. This happens because a preferential flow path is set up. Soil permeability decreases with depth, meaning that the downward movement of water is slowed. During rainfall, this leads to a backing up effect in the more permeable surface layers. This has been likened to the flow of water down a thatched roof: it is easier for the water to move parallel to the slope of the roof, along the stems of the straw, than to move vertically downwards through it (Ward, 1984; Zaslavsky and Sinai, 1981).

Where they exist, **soil pipes** provide a very rapid throughflow mechanism, and rates of flow can be comparable to those in surface channels. Soil pipes are hydraulically formed conduits that can be up to a metre or more in diameter. They are found in a wide range of environments and are sometimes several hundreds of metres

in length (Jones, 1997). As such, they can act as an extension to the channel network, allowing the drainage basin to respond rapidly to precipitation inputs (Jones, 1979).

The schematic diagram in Figure 3.1 represents the headwaters of a humid zone river, where the channel intersects the water table surface (it should be noted that aquifers are not always present, for example aquifer development is extremely limited in headwater areas that are underlain by impermeable rocks and characterised by thin soils). In this example, groundwater contributions are made to the channel flow from the underlying aquifer and the channel is called a **gaining stream**. Even during rainless periods, flow will be maintained, as long as the level of the water table does not fall below that of the channel bed. A rather different situation exists for many dryland channels, where the water table may be several metres, or tens of metres, below the surface. In this case, the direction of flow is reversed, as water is lost through the bed and banks of the channel, percolating downwards to recharge the aquifer. This is termed a **losing stream** and, although not exclusive to drylands, many examples are found in these environments. It is not unusual for a river to be gaining and losing flow to groundwater along different parts of its course, while seasonal fluctuations in water table levels can mean that losing streams become gaining streams for part of the year. The Euphrates in Iraq provides a good example, with most of the flow being generated in the headwaters in northern Iraq, Turkey and Syria. Further downstream, at the Hit gauging station (150 km west of Baghdad) the river loses flow to groundwater for much of the year (Wilson, 1990). The loss of flow from a channel, due to downward percolation and high evaporation rates, is referred to as **transmission loss**.

The storm hydrograph and drainage basin response

Flow discharge (also known as Q) is the volume of water passing through a given channel cross-section in a given time. The units of discharge most commonly used are cubic metres per second ($m^3 s^{-1}$), known as 'cumecs', although for very small flows litres per second may be used. In the United States cubic feet per second, or 'cusecs', are used instead of cumecs. Box 3.1 explains how discharge is monitored.

Box 3.1

THE MEASUREMENT OF STREAMFLOW

The channel discharge is the volume of water flowing through a given channel cross-section in a given time. A number of different methods have been developed to measure discharge. These can be grouped into **instantaneous measurements**, where discharge is measured at a particular point in time, and **continuous measurements** for a record of discharge variations through time.

The velocity–area method

Discharge is measured in cubic metres per second (m^3s^{-1}). It increases with the area of the channel cross section and with the velocity of flow. Discharge can be calculated for a given channel cross-section by measuring its cross-sectional area and the mean flow velocity:

$$Q = A \times v$$

where Q = discharge, A = cross-sectional area and v = mean flow velocity.

Figure 1(a) illustrates the method used. The first stage is to stretch a tape measure across the width of the channel. The channel is then divided into a number of sub-sections. This varies according to the width of the river, but discharge is usually gauged at twenty or more subsections. Ideally, the discharge flowing through each sub-section should be similar, so these are more closely spaced where the flow is deeper and faster.

Velocity measurements are then made using a flow meter. A commonly used design is a propeller mounted on a rod, which is lowered into the flow. The flow velocity is directly proportional to the rate at which the propeller is turned by the flow. A digital readout shows the velocity in m s^{-1} or, if you are not so lucky, the number of rotations per minute. (This can then be converted to the equivalent velocity using a simple formula.) One velocity measurement is made for each of the sub-sections, at the points indicated in Figure 1(a). Since flow velocity increases from zero at the channel bed to a maximum near the water surface,

a representative average flow needs to be measured. It can be shown that the flow velocity at a height of $0.4d$ above the bed is representative of the average flow velocity, where d is the total depth of flow. Velocity measurements should therefore be made at a distance below the water surface that is 0.6 of the total depth (i.e. 0.4 of the depth from the bed). Thus if the depth of flow was 1 m, you would measure the velocity at a depth of 0.6 m below the surface (or 0.4 m above the bed).

The width and depth of each sub-section are measured as shown in the diagram. The discharge flowing through each sub-section can be calculated by multiplying the sub-section width (w), depth (d) and velocity (v). In order to calculate the discharge for the whole cross-section, the discharges for each of the sub-sections are added together. The discharge flowing through the 'left over' triangles adjacent to each bank is usually assumed to be negligible and is not included in this calculation.

This method is not very suitable for steep, turbulent, rocky streams where accurate current meter measurements are hard to obtain. A more appropriate technique in this case is to use **dilution gauging**. A chemical 'tracer' substance such as salt or dye is released into the flow, either as a single large 'gulp' or 'slug', or by continuously injecting it into the flow. Changes in concentration are monitored further downstream. Since the amount of dilution increases with discharge, it is possible to relate the change in concentration to stream discharge.

Continuous streamflow measurement

Although laborious, the velocity-area method only gives you one measurement of discharge at a particular point in time. In order to plot hydrographs like the ones in Figure 3.2, it is necessary to have a continuous record of flow.

Discharge is difficult to measure directly. However, it is related to **stage**, the water level or height, which is much easier to record. Figure 1(b) shows a gauging

Box 3.1

THE MEASUREMENT OF STREAMFLOW—CONT'D

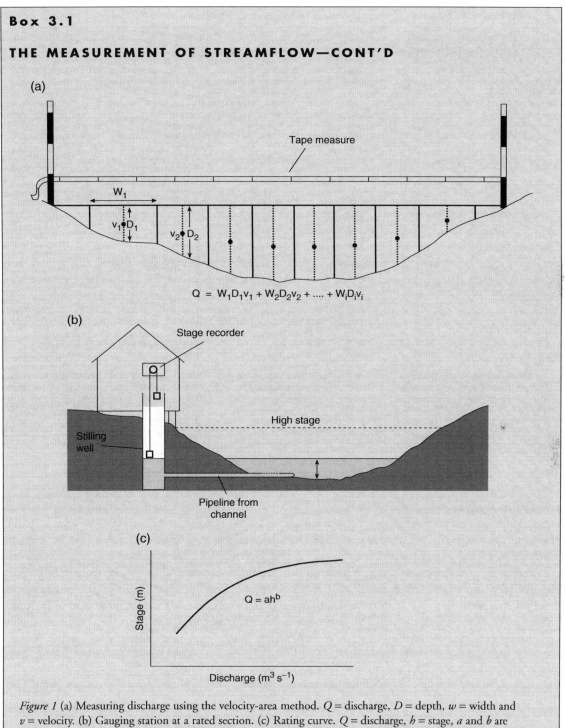

$$Q = W_1D_1v_1 + W_2D_2v_2 + ... + W_iD_iv_i$$

Figure 1 (a) Measuring discharge using the velocity-area method. Q = discharge, D = depth, w = width and v = velocity. (b) Gauging station at a rated section. (c) Rating curve. Q = discharge, h = stage, a and b are coefficients.

(Continued)

Box 3.1

THE MEASUREMENT OF STREAMFLOW—CONT'D

station situated on a natural cross-section. Rather than measuring the water level in the river itself, a pipeline leads to a **stilling well** which damps out the effects of surface waves and turbulence. The level of water in the stilling well is monitored using a **stage recorder**. Traditionally this would have consisted of a float attached to a counterweight via a pulley (shown in Figure 1b). As the water level rises and falls the float moves up and down, turning the pulley. This is attached to a pen, which draws a trace on a chart mounted on a rotating drum. The disadvantage is that the charts then have to be read manually. Digital measurements of water level can be made by sending an infrared signal down the stilling well. This is reflected by the water surface back to a receiver, with the time delay measured.

Stage is converted to discharge using a **rating curve** (Figure 1c). Using this, the discharge corresponding to a given stage can either be read off from the graph or calculated using the rating equation. This is of the form $Q = ah^b$, where Q = discharge, h = stage and a and b are coefficients, which describe the unique relationship between stage and discharge for the cross-section.

This relationship is affected by the shape of the cross-section. At lower flows, small increases in discharge result in relatively large increases in stage because the channel is narrow near its bed. As the flow increases, a greater increase in discharge is needed to produce the same increase in depth, because the channel widens above the bed. Therefore a wide, shallow channel will have a different relationship (relatively small increases in stage with rising discharge) to a narrow, deep channel (relatively large increases in stage with rising discharge).

The rating curve is derived from discharge measurements made at different stages. Measurements are difficult to make at high flows, so most rating curves tend to be less reliable at high flows. Another problem is that if a large flood alters the shape of the cross-section, the rating curve has to be re-calibrated. This can be avoided using a gauging structure such as a **flume** or **weir**. These have a regular cross-sectional area and the flow velocity is controlled by the structure. The rating curve can be derived using a combination of measurements and hydraulic theory.

The top graph in Figure 3.2 (solid line) shows an **annual hydrograph** of daily flows, which might be observed for a river in the temperate zone over the course of a year. The lower graph is a single **storm hydrograph**, which shows the response of the drainage basin to one precipitation event. During a particular rainfall event, there is a delay between the onset of rainfall and the time at which the discharge starts to increase. The initial increase is due to water falling directly into the channel and close to it, though as the storm progresses, water travelling from greater distances reaches the channel. Water taking the fastest pathways – overland flow, shallow throughflow and pipe flow – is rapidly transferred to the channel, contributing to the **quick flow** component of the storm hydrograph. **Base flow** contributions come

from water taking the slower subsurface routes, taking much longer to reach the channel. This means that water continues to enter the channel as base flow for some time after rainfall has ceased, keeping rivers flowing during dry periods. Glaciers, lakes, reservoirs and wetlands also contribute to base flow. The relative proportions of quick flow and base flow determine the size of the hydrograph peak and the time delay, or **lag time**, between peak rainfall and peak flow. For instance, where quick flow dominates, lag times are relatively short and peak flows relatively high. Rivers dominated by base flow respond more slowly and the peak flow is lower. The two lines shown in Figure 3.2(a) represent two temperate zone rivers, identical in every respect apart from the underlying geology: basin 1 is underlain by impermeable rocks

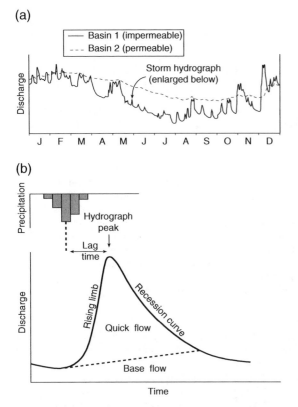

(a)

(b)

Figure 3.2 (a) Typical annual hydrograph of a temperate zone river whose drainage basin is underlain by an impermeable geology (solid line). Also shown is the annual hydrograph of a basin with a more permeable geology (dotted line). (b) Characteristics of a storm hydrograph.

and basin 2 by permeable rock. From the graphs you will see that basin 1 has a flashy response, with marked storm peaks, whereas basin 2 has a more damped response. This is because most of the precipitation falling over basin 2 infiltrates and takes slower subsurface pathways to the stream channel. There is also a marked difference in the summer low flows, with the greater base flow component for basin 2, which sustains higher summer flows.

The hydrological response of a river to discrete inputs of precipitation through time, as indicated by the shape of its hydrograph, is determined by the drainage basin characteristics and climatic factors shown in Table 3.1.

ANNUAL FLOW REGIMES

The **annual flow regime** of a river describes the seasonal variations in flow that are observed during an 'average' year. As you might expect, this is influenced by the seasonal distribution of rainfall, and the balance between rainfall and evaporation at different times of year. For example, some tropical rivers experience a marked wet and dry season, drying up completely for part of the year and carrying high flows during the wet season. Climate also has an important influence on the type and density of vegetation, soils and land use, all of which act as controls on the processes of runoff generation (Table 3.1).

Several climate characteristics are important in determining the flow regime. These include whether it is humid or arid, if it is predominantly warm or cold, the annual range of temperatures, and whether precipitation is seasonal or occurs all year round. At high latitudes and in some mountain environments, the timing and length of glacial ablation and snowmelt is a dominant factor. Figure 3.3 shows a selection of typical flow regimes, which characterise different climatic zones. These come from a classification scheme developed by Beckinsale (1969) from an existing climate classification. The different regimes are categorized using a system of two letters. The first letter relates to the mean annual precipitation and annual temperature range:

A: Warm, moist tropical climates, where the mean temperature exceeds 18°C for all months of the year.

B: Dry climates, where rates of potential evaporation[1] exceed annual precipitation.

C: Warm moist temperate climates.

D: Seasonally cold climates with snowfall, where the mean temperature is less than –3°C during the coldest month.

The second letter indicates the seasonal distribution of precipitation:

F: Appreciable rainfall all year round.

W: Marked winter low flow.

S: Marked summer low flow.

For example, the regime of the Pendari River (Figure 3.3A) is influenced by a tropical climate with a marked winter low flow, and would be classified as AW.

Table 3.1 Factors affecting hydrological response of a basin

Soils and geology

- *Soil type and thickness.* Soil texture (relative proportion of sand, silt and clay particles) affects infiltration rates. Sandy soils have high permeability whereas clay soils do not. In arid areas a crust can form on the soil surface, decreasing the permeability. Soil thickness affects how much water the soil can absorb.
- *Geology.* Drainage basins underlain by a permeable geology tend to have a slower response to precipitation, although the flow is sustained for a longer time during dry periods. Drainage basins underlain by impermeable materials have a faster, or more 'flashy' response.

Vegetation and land use

- *Vegetation type and density.* Vegetation reduces the impact of raindrops and allows a more 'open' soil structure, meaning that infiltration rates are higher. Vegetation also affects interception rates and evapotranspiration losses from the basin.
- *Urban areas.* Depends on the proportion of the drainage basin that is urbanised. Large areas of paved surfaces, drains and culverts rapidly transmit water to river channels, leading to an increase in peak flow and a shorter lag time.
- *Grazing and cultivation.* When deforestation occurs, rates of overland flow tend to increase. Heavy machinery and trampling by animals compact the soil, reducing permeability, although ploughing can increase infiltration rates. Flow may be concentrated in plough furrows that run up and down the slope.
- *Land drainage.* The installation of field drains allows rapid transfer of runoff into the nearest stream channel.

Physiographic characteristics

- *Drainage basin size and shape.* In larger basins the travel times are longer, as flow has to travel greater distances to reach the outlet. The total volume of runoff increases with the drainage area. Elongated drainage basins have a response that is initially more rapid but with a lower, more gentle peak.
- *Drainage density.* Where the density of stream channels is high, the average distance over which water has to travel to reach the channel network is reduced, leading to a more rapid response.
- *Drainage basin topography.* Travel times are increased over steep slopes. In upland areas, steep slopes are often associated with thin soils and the response tends to be flashy. Rainfall may be affected by altitude and aspect with respect to storm tracks.

Channel characteristics

- *Channel and floodplain resistance.* The velocity of flow in river channels is affected by the roughness of the bed and banks and the shape of the channel. Overbank flows are slowed by the roughness of the floodplain surface.
- *Floodplain storage.* When channel capacity is exceeded, water spills out onto the surrounding floodplain, where it is stored until the floodwaters recede. If floodplain storage is limited, a greater volume of water travels downstream.
- *Conveyance losses.* In dryland environments the channel may lose flow due to high rates of evaporation and 'leakage' by exfiltration through the channel boundary.

Meteorological factors

- *Antecedent conditions.* The conditions in the drainage basin prior to the onset of precipitation. Where recent or prolonged previous rainfall has occurred, the soil may be near saturation, meaning that a relatively small input of rainfall could lead to a rapid runoff response. Where snow is lying on the ground, subsequent rainfall can cause it to melt, which may lead to flooding downstream.
- *Rainfall intensity.* Rainfall intensity is expressed in millimetres per hour (mm h^{-1}). The more intense the rainfall, the more likely it is that the infiltration capacity of the soil will be exceeded.
- *Rainfall duration.* This is the period of time over which a given rainfall event takes place. As the storm progresses, runoff contributing areas at greater and greater distances from the channel network become active. The channel network may also extend upstream as normally dry channels start to carry flow.

The Arno (Figure 3.3C) has a warm temperate rainy climate, with summer low flows, and is classified as CS.

On first looking at Figure 3.3, it might seem strange that the graph for dry climates (B) has the biggest peak. Bear in mind that these graphs indicate annual flow variability, rather than actual monthly discharges. The value for each month is the ratio of the monthly mean flow to the overall (annual) mean, which is shown by the dotted line on each graph. Regimes associated with **tropical, rainy climates** (Figure 3.3A) are affected by seasonal shifts in the inter-tropical convergence zone. Near the equator, runoff occurs year-round, with peaks at the equinoxes (Lobaye River). Further north and south, there are marked wet and dry seasons (Pendjari River). The annual runoff for rivers in **dry climates** (Figure 3.3B) is low, but extremely variable. For much of the time there is little or no flow, but extreme floods can also occur. Flood pulses on the Cooper Creek are highly erratic and do not occur every year, since they are associated with El Niño disturbances to the monsoon. In contrast, there is much less variability for rivers with **temperate rainy climates** (Figure 3.3C). These have higher flows in winter and relatively low flows in summer, a pattern which is accentuated in Mediterranean climates (Arno). Snowmelt peaks are seen for rivers with **seasonally cold, snowy climates** (Figure 3.3D). The timing of the peak is dependent on altitude, latitude and seasonal patterns of rainfall. In **high mountain environments**

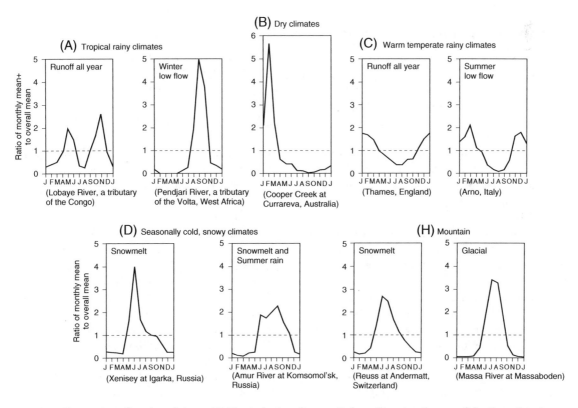

Figure 3.3 Flow regimes for selected rivers. (A) Tropical rainy climates: Lobaye River (a tributary of the Congo) and Pendjari River (a tributary of the Volta in West Africa). (B) Dry climates: Cooper Creek at Currareva, Australia. (C) Warm, temperate, rainy climates: Thames, England, and Arno, Italy. (D) Seasonally cold, snowy climates: Xenisey at Igarka and Amur River at Komsomol'sk, both in Russia. (H) Mountain climates: Reuse at Andermatt and Mass River at Massaboden, both in Switzerland. Source: Adapted from Beckinsale (1969); (B) after Knighton and Nanson (1997).

(Figure 3.3H), class HN denotes drainage basins with a nival regime (where snow patches exist) and HG those with a glacial regime. H stands for *Höhenklima*, which translates from the German as 'highland climate'.

For any drainage basin, inter-annual variations will occur between wet and dry years, so the flow regime is described using long-term averages. Even within a particular climate zone, the flow regime can differ markedly between drainage basins. This is because the physical characteristics of a drainage basin also play an important role, most notably the underlying geology and soil characteristics. These determine how much storage is available within the drainage basin in natural reservoirs, such as groundwater, lakes and wetlands.

Downstream variations in discharge

As well as varying through time, discharge also changes along the course of a river. At any location, channel form is dependent on the discharge and supply of sediment from upstream. In most cases, discharge increases downstream as the area of the drainage basin increases and tributaries join the main channel. There is also a general increase in the size of the channel, with discharge acting as a control on the gross dimensions (Knighton, 1998). The quantitative description and understanding of the nature of these downstream changes have been the focus of much research and is explored further in Chapter 8. Although there is a *general* downstream increase in channel dimensions, local influences lead to considerable variation, even over short distances.

Downstream changes in dryland channels can be very pronounced (Tooth, 2000). For example, infrequent flooding occurs along the ephemeral streams draining the Barrier Range in arid western New South Wales, Australia. Away from the uplands, high transmission losses lead to a rapid downstream reduction in discharge and channel size (Dunkerley, 1992). Downstream reductions in discharge and cross-sectional area are also observed in the piedmont and lowland zones of rivers draining the northern plains of arid central Australia (Tooth, 2000). There is an eventual termination of channel flow and bedload transport. However, during large floods, flows continue out across extensive unchannelled surfaces called 'floodouts' (Tooth, 2000). Tooth highlights the complex interactions between discharge, sediment transport, channel slope, tributary inflows, bank sediments and vegetation. These give rise to considerable variations in the downstream channel changes observed for dryland rivers.

FLOODS

Although the flow regime shows seasonal variations in river flow, it does not provide detailed information on the magnitude (size) and frequency of floods and droughts. Floods are of most interest here because they are capable of carrying out large amounts of geomorphological work and are thus significant in shaping the channel.

The term 'flood' is hard to define. In general terms, a flood is a relatively high flow that exceeds the capacity of the channel. While more frequent flows are confined within the channel, periodic high flows overtop the banks and spill out onto the surrounding floodplain. Significant here is the **bankfull discharge** (Q_b), defined as 'that discharge at which the channel is completely full' (Knighton, 1998). Although these definitions may sound straightforward enough, it is actually quite difficult to define bankfull discharge in the field because the height of the banks varies, even over short distances. This means that overtopping of the banks does not occur simultaneously at all points along the channel. Floodplain relief can be quite variable, with variations of between 1.7 m and 3.3 m observed on three Welsh floodplains (Lewin and Manton, 1975). Along the Alabama River, United States, flooding has been observed to occur more frequently at the apexes of actively migrating meander bends. This is associated with the development of floodplain features called levees. These are raised ridges that form along the banks when material is deposited during overbank flows (see Figure 8.9). Levee development is impeded at actively migrating bends because the deposits are eroded as the channel migrates. Levees are better developed (higher) along less actively migrating sections of channel, where flooding occurs less frequently (Harvey and Schumm, 1994).

Box 3.2

CALCULATING FLOOD RETURN PERIODS

Flood return period should ideally be calculated on the basis of at least thirty years worth of flow data. If possible a longer record should be used as this will contain a larger number of flood events and will provide a more representative sample of all flood events. The first step is to identify the peak flow for each year in the record to produce an **annual maximum series**. The **mean annual flood** is given by the mean of this annual maximum series. Mathematical analyses have shown that the recurrence interval of the mean annual flood is 2.33 years (Leopold *et al.*, 1964). In other words, this flow will be exceeded by the highest flow of the year once every 2.33 years on average.

All the floods in the annual maximum series are then ranked in order of magnitude, with the largest event ranked first and progressively smaller events given higher numbers. A simple formula is then used to assess the return period in years:

$$T = \frac{(n+1)}{m}$$

where T = return period in years, n = rank and m = number of years in record

If this is plotted using a logarithmic scale for flood magnitude the flood frequency curve is transformed to a straight line. It is possible to extrapolate (extend) this line to estimate the size of floods with larger return periods than those on record, although the difficulties associated with fitting a best-fit line through the existing data mean that a small difference in its gradient could make a big difference to the estimated size of the flood. There are also a number of practical difficulties associated such estimates and errors of flood discharges and errors in flood discharge estimates are generally considered to be in the range of 10 per cent to 100 per cent (Benito *et al.*, 2004). Large floods are very difficult to record accurately, since gauging stations can be damaged or even destroyed, leading to critical gaps in the flood record. Estimates of flood discharges are also dependent on the quality of the rating curve (relationship between stage and discharge – see Box 3.1).

Flood magnitude and frequency

Floods of different sizes are defined in terms of high water levels or discharges that exceed certain arbitrary limits. The height of the water level in a river is called its **stage**. For a given river, there is a relationship between the size of a flood (in terms of its maximum stage or discharge) and the frequency with which it occurs. Floods of different sizes do not occur with the same regularity: large floods are rarer than smaller floods. In other words, the larger the flood, the less often it can be expected to occur. Floods are therefore defined in terms of their magnitude (size) and frequency (how often a flood of a given size can be expected to occur).

You have probably heard reference to the 'twenty-year flood' or the '100-year flood'. This **return period** is an estimate of how often a flood of a given size can be expected to occur and, since less frequent floods are more extreme, the 100-year event would be bigger than the twenty-year flood.

The return period (T) can also be expressed as a probability (P) by taking the inverse of the return period, i.e.:

$$P = \frac{1}{T}$$

Using this, the probability of a 100-year flood taking place in any one year can be calculated as

0.01 (i.e. 1 per cent), and for the twenty-year flood, 0.05 (5 per cent). The probability that a flood with a particular return period will occur is the same every year and does not depend how long it was since a flood of this size last occurred – the twenty-year does not occur like clockwork every twenty years. However, if a period of several years is considered, the likelihood of a given flood occurring during this time increases. For example, if someone bought a house on the 100 year floodplain and lived there for thirty years, the probability of that property being flooded in any one year would be 0.01. This increases to 0.3 (probability × number of years), or 30 per cent, for the thirty-year period. Box 3.2 explains how return periods are estimated. As with any odds, flood probabilities are *estimates*, and a number of underlying assumptions are made when deriving them. It is assumed that runoff is randomly distributed through time and that the data set holds a representative sample of these random events. Estimates are therefore more reliable when a longer record is available, since a larger number of flood events will be included in it. Another assumption is that there are no long-term trends in the data, which is not the case when climate change is occurring.

The frequency of bankfull discharge

Although bedrock channels are mainly influenced by high magnitude flows, those formed in alluvium can be adjusted by a much greater range of flows (see Chapter 1, pp. 5–6). This is reflected by the morphology and size of alluvial channels. Over the years, much research has focused on the bankfull discharge (defined above), since it represents a distinct morphological discontinuity between in-bank and overbank flows. Leopold and Wolman (1957) suggested that the channel cross-section is adjusted to accommodate a discharge that recurs with a certain return period. From an examination of active floodplain rivers, they found that the bankfull discharge had a return period of between one and two years. This is corroborated by later observations made for stable alluvial rivers (for example, Andrews, 1980; Carling, 1988). However, the concept of a universal return period for bankfull discharge that can be applied to all rivers is controversial. Williams (1978) observed wide variations in the frequency

of bankfull discharge, which ranged from 1.01 to 32 years, and concluded that this was too variable to assume a uniform return period for all rivers. Even along the same river, there can be marked variations in the frequency of bankfull discharge (Pickup and Warner, 1976).

The concept of a uniform frequency for bankfull discharge assumes that all channels are '**in regime**'. This means that the morphological characteristics of a given channel, such as size, fluctuate around a mean condition over the time scale considered (Pickup and Reiger, 1979). This is not true for all rivers and there are many examples of **non-regime**, or **disequilibrium**, channels. An example would be where channel incision is taking place through erosion of the channel bed. This results in a deeper channel, which requires a larger, and therefore less frequent, discharge to fill it. The Gila River in Arizona, United States, was greatly enlarged when past events had led to large floods. The enlarged channel is not adjusted to the contemporary flow regime, which means that the bankfull discharge for the enlarged channel has a much lower frequency (Stevens *et al.*, 1975). The material forming the bed and banks is also significant. In cases where the boundary is very erodible, the bankfull discharge may simply reflect the most recent flood event (Pickup and Warner, 1976).

The geomorphological effectivenes of floods

Given that many rivers exceed their channel capacity and flood on a fairly regular basis, it would not be unreasonable to ask why they do not shape channels that are large enough to convey all the flows supplied to them. While it is true that high-magnitude events lead to significant changes in channel morphology, the comparative rarity of these large floods must also be taken into account. The cumulative effect of smaller, more frequent floods can also be significant in shaping the channel. The effectiveness of any given discharge over a period of time is therefore something of a compromise between its size and how often it occurs. The basic question is: are a number of smaller floods as effective as one large flood? This concept is explored further in Box 3.3.

Box 3.3

THE FREQUENCY AND MAGNITUDE OF CHANNEL FORMING FLOWS

For a given channel, the geomorphological effectiveness of a particular flood event is dependent on the magnitude or size of that flood. This is because larger floods have more potential to erode and transport sediment. However, smaller floods, although less effective on an individual basis, occur more frequently. Over a period of time – decades to centuries – the cumulative volume of sediment transported by a number of these smaller floods can be greater than for one or two major floods (Wolman and Miller, 1960).

This concept is represented by Wolman and Miller's graphs, which are shown in Figure 1. The discharge frequency curve shows the frequency distribution of different flows ranging from droughts and low flows (left hand side) to large floods (right hand side). It can be seen that normal (non-flood) flows prevail for most of the time, while low flows and floods occur less often. The frequency of a given flood decreases with its magnitude.

The sediment transport rate curve represents the volume of sediment transported by individual floods of a given magnitude. This shows that the sediment transport rate increases with flood magnitude. The third curve, shown as a dotted line, indicates the cumulative effectiveness of a given flow over time. It shows the product of the sediment transport rate for that flow and its frequency of occurrence. The indication is that moderate, high frequency floods are most geomorphologically effective.

It has been suggested that channel parameters such as the spacing of meander bends, channel width and bankfull discharge are scaled to a single, **dominant discharge**. Indeed the frequency of bankfull discharge

is similar to the flow that cumulatively transports the greatest volume of sediment. At least, that is *generally* the case for alluvial channels in humid temperate regions, where bankfull discharge typically occurs every one to two years.

However, despite the attractiveness of such a theory in understanding and modelling the adjustment of channel form, it does have serious limitations. For example, many channel geomorphic units such as bars and bedforms are adjusted by normal flows. Heritage *et al.* (2001), working on mixed bedrock–alluvial channels, found that while channel dimensions are scaled to flood flows, features such as bars within the channel are adjusted to lower flows. Thus, rather than being adjusted to a single, 'dominant' discharge, channels tend to adjust to a range of flows.

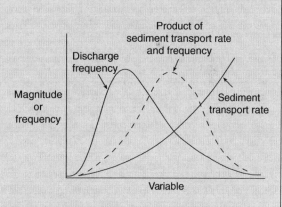

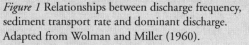

Figure 1 Relationships between discharge frequency, sediment transport rate and dominant discharge. Adapted from Wolman and Miller (1960).

Regional flood frequency curves

The flood frequency–magnitude relationship differs between regions. Despite the low annual rainfall in dry-land environments, precipitation can be highly variable and the twelve largest floods ever recorded in the United States all occurred in semi-arid or arid areas (Costa, 1987). During flash floods, such as the one shown in Colour Plate 14, floodwaters rapidly inundate the dry channel. Not all dryland rivers are prone to flash flooding however, and there is considerable variation in the size, type and duration of flooding.

Regional flood frequency curves are shown in Figure 3.4. The return period is plotted on the horizontal axis using a logarithmic scale, with the relative flood magnitude on the vertical axis. A *relative* flood magnitude has been used to allow comparison between floods for a number of rivers in different regions. Because these all drain different areas, a direct comparison of flood magnitudes would not be very meaningful. Instead, for each river included in the analysis, the ratio between the magnitude of each flood on record and a low magnitude 'reference flow' – the mean annual flood – has been used. This is defined in Box 3.2 and has a return period of 2.33 years (i.e. the flow that will be equalled or exceeded on average once every 2.33 years[2]). The steepness of each curve reflects the variability of the flow, with arid zone rivers showing a much greater increase in relative flood magnitude at higher return periods. This reflects the extreme flow variability observed in these rivers and has important implications for the morphology of dryland channels, as will be seen in later chapters.

Reconstructing past floods

Palaeoflood hydrology is a new and developing area of hydrology and geomorphology, which reconstructs past flood events in order to extend the flow record. Due to problems associated with monitoring major floods and the relatively short duration of most gauged records, extreme floods are very rare in the observational record. By reconstructing palaeofloods, the flood record can be extended, allowing increased accuracy in the estimation of floods for risk analysis (Box 3.2). Evidence of past

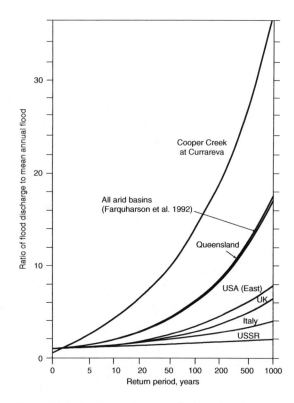

Figure 3.4 Regional growth curves, showing the relation between flood magnitude and frequency for various regions. For selected flood events the flood magnitude, relative to the mean annual flood, is plotted against the return period of that flood. After Knighton and Nanson (1997).

flood events is provided by geological indicators such as flood deposits, silt lines and erosion lines along the channel and valley walls (Benito *et al.*, 2004). Historical records are also used and include documents, chronicles and flood marks inscribed on bridges and buildings. Using this evidence, it is possible to determine the size of the largest flood events over periods of time ranging from decades to thousands of years (Benito *et al.*, 2004). As well as identifying the largest floods, evidence of floods above or below above specified flow stages can also be reconstructed (Stedinger and Baker, 1987). Although time-consuming, it is possible to reconstruct a complete record, chronicling the largest flood, together with the size and number of intermediate palaeofloods (Benito *et al.*, 2004). Chapter 9 discusses some of the techniques that are used in reconstructing past flood events.

CHAPTER SUMMARY

Inputs of precipitation falling over the area of a drainage basin are transferred to the channel via a number of different pathways. These include surface overland flow, throughflow (through the soil) and deeper groundwater flow. Rates of movement vary considerably: overland flow and shallow throughflow are generally much more rapid than groundwater flow. An important control is the infiltration capacity, which determines how quickly water can be absorbed by the soil. If the rainfall intensity is greater than the infiltration capacity, excess water builds up at the surface, leading to overland flow. Overland flow also occurs when rain falls on saturated areas. The volume of water flowing through a given cross-section in a given time (discharge) fluctuates in response to inputs of precipitation. A hydrograph shows how discharge changes over the course of a year (annual hydrograph), or one rainfall event (storm hydrograph). The shape of the hydrograph is affected by the physical, land use and climatic characteristics of the drainage basin. These variables all determine the relative proportion of water taking faster and slower pathways to reach the channel. Climate is a very important control on the annual flow regime of a river, which reflects the precipitation amount, seasonal distribution and annual temperature variations. Another important characteristic of the flow regime is the frequency and magnitude (size) of flood events. As the size of a flood increases, the frequency with which it occurs (return period) decreases. The relationship between frequency and magnitude differs from region to region. In dryland environments, large, low frequency floods are much more extreme than those with a similar return period in humid areas. The bankfull discharge is that flow at which the channel is completely filled. Wide variations are seen in the frequency with which the bankfull discharge occurs, although it *generally* has a return period of one to two years for many stable alluvial rivers. The geomorphological work carried out by a given flow depends not only on its size but also on its frequency of occurrence over a given period of time.

FURTHER READING

Introductory texts

Davie, T., 2003. *Fundamentals of Hydrology*. Routledge, London. Provides more detail on runoff generation and hydrograph analysis without getting too technical.

Jones, J.A.A., 1997. *Global Hydrology*. Longman, Harlow. Good on processes of runoff generation, measurement, and the effects of land use change.

Manning, J.C., 1997. *Applied Principles of Hydrology*. Third edition. Prentice Hall, Englewood Cliffs NJ. Written for students with a limited science background, this provides a very readable and accessible introduction.

Ward, R.C. and Robinson, M., 1990. *Principles of Hydrology*. McGraw-Hill, London. A well established hydrology textbook, which includes an interesting discussion on the development of runoff generation theory.

Websites

Dryland Rivers Research website, http://www. drylandrivers.com. An informative web site that covers many aspects of dryland rivers, with definitions, images, features, new research and useful links.

Hyperlinks in Hydrology for Europe and the Wider World, http://www.nerc-wallingford.ac.uk/ih/devel/wmo. Numerous links to websites of organisations active in various aspects of hydrology, including many educational sites.

CEH Wallingford, http://www.ceh.ac.uk/data/index. html. National River Flow Archive for the UK. Time series download facility (daily flows), hydrological summaries.

Global River Discharge Database, http://www.sage.wisc.edu/riverdata/index.php?qual=32. A compilation of monthly mean discharge data for over 3,500 sites worldwide.

US Geological Survey, http://water.usgs.gov/osw/. Includes an extensive database, real-time flow data, current information on flooding and drought in the United States, and education information on many aspects of hydrology.

4

SEDIMENT SOURCES

This chapter examines the initial production of the sediment that is supplied to, and transported by, river channels. Much of this sediment originates in the production zone of the headwater regions (see Figure 2.1). Sediment is originally derived from the underlying bedrock which, having been broken down by weathering, is then transported down-slope by processes of mass movement and water erosion. River bank erosion also produces a significant proportion of the sediment transported in river channels. In glaciated drainage basins, vast quantities of sediment are produced by glacial erosion. Aeolian (wind blown) material may also provide a source of sediment. Figure 4.1 provides a generalised schematic illustration of the main sediment sources. It should be noted that the relative dominance of these various processes varies considerably across different environmental settings.

Inputs of sediment to the channel network are highly variable over both space and time. There is a stochastic, or random, element to many sediment generation processes, which can be initiated or greatly accelerated by the occurrence of rainstorms, earthquakes and wildfire. As with flood events (Chapter 3), rainstorms can be defined in terms of a probability distribution. Important characteristics include rainfall intensity, duration, the length of time between events and the spatial extent of a given storm. The effects of a given

rainstorm are influenced by spatial variations in the landscape on which it occurs. These include topography, soil properties, vegetation cover and land use, all of which lead to spatial differences that can be seen in the operation of processes over a wide range of scales. As a result, sediment influxes to the channel network occur as a complex sequence of pulses. Only a certain proportion of the sediment produced by these processes actually reaches the channel network. This is dependent on

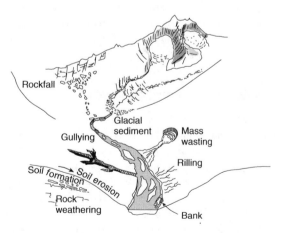

Figure 4.1 Schematic illustration of different sediment sources.

the degree of connectivity that exists between the sediment source and the river channel, which will be discussed in Chapter 5.

In this chapter you will learn about:

- The breakdown of rocks by processes of weathering.
- Down-slope transport by different types of mass movement processes.
- The erosion of the soil surface by the action of raindrops, overland flow and concentrated flow in rills and gullies.
- How human activity can lead to accelerated soil erosion and the management problems that are associated with it.
- Methods for assessing rates of soil erosion.

WEATHERING

Weathering involves the physical breakdown and chemical decomposition of rocks and loose rock material *in situ* at the Earth's surface. It takes place when these parent materials are exposed to environmental conditions at the Earth's surface, where many rocks become unstable. This is because the conditions there may be very different to those under which they were formed, perhaps at high temperatures and pressures some depth below the surface.

Weathering produces a primary source of sediment for removal by processes of erosion and mass wasting. It also provides mineral matter as a raw material in soil formation. Here we are mainly interested in weathering as a source of sediment, so the actual processes will not be considered in detail.

Physical weathering processes

Weathering processes may be classified as physical or chemical, although this is a theoretical distinction because in most cases physical and chemical weathering act together. Physical weathering involves the mechanical breakdown of rocks into smaller fragments without any change taking place in their chemical condition. Processes include breakdown due to stresses exerted by the growth of crystals of ice or salt in rock crevices; pressure release caused by the denudation and removal of layers of overlying rocks; and thermal expansion and contraction

caused by heating and cooling due to wildfire or solar radiation. In mountainous regions, where the temperature regularly fluctuates above and below zero, **freeze–thaw action** is a very effective breakdown process. Water expands when it freezes and the repeated formation and melting of ice crystals can exert considerable stresses.

Chemical weathering processes

Chemical weathering brings about changes in the chemical, mineralogical and physical properties of rocks. Alteration of the constituent minerals often leads to a weakening of the overall structure. For example, many of the secondary minerals formed by the breakdown of primary minerals in rocks are less dense but take up a greater volume. The most important group of secondary minerals are the clay minerals.

Hydrolysis is one of the main chemical weathering processes. It can drastically modify and decompose susceptible primary minerals, and is significant in the conversion of parent material to clay. Hydrolysis involves a reaction between water and silicate minerals in rocks and soil in which mineral cations such as potassium, calcium and magnesium are replaced by hydrogen ions. This weakens the internal structure and leads to breakdown.

Carbonation dominates the weathering of calcareous rocks, such as limestone and dolomite. It is a complex process, brought about when carbon dioxide from the atmosphere is dissolved in rainwater to form a weak carbonic acid. This is able to attack calcareous rocks by forming water-soluble carbonates. Carbonation is a step in the complex weathering of many other minerals, such as the hydrolysis of feldspar (Huggett, 2003).

Water is also involved in the more 'mechanical' processes of **solution**, where mineral salts dissolve in this very effective solvent, and **hydration**, where the water molecules are absorbed by certain minerals but do not alter the minerals chemically.

Weathering reactions involving oxygen also occur. Iron-bearing minerals are susceptible to **oxidation** where oxygen dissolved in soil waters combines with minerals within the parent rock. This alters the crystal lattice and makes it more prone to further breakdown (Huggett, 2003).

Weathering products

The surface of exposed bedrock is often mantled by a layer of weathered material called **regolith**. This includes solid particles, ranging in size from large boulders to fine clay particles, material carried away in solution and colloids (microscopic particles).

The thickness of the weathering mantle is determined mainly by the balance between rates of weathering, and the relative rate at which material is transferred away from the site by mass movement and agents of erosion (water, wind and ice).

Rates and controls

Rates and types of weathering are determined by four main groups of variables: climate, parent material, topography and organic activity. These operate over different scales. For example there is a reasonable correlation between the major climatic regions and different weathering zones, while the effects of topography and parent material tend to be more localised.

The two climate-related parameters of greatest significance are temperature and precipitation, the latter being the primary control on water availability. Rates of chemical weathering increase with temperature, approximately doubling for every 10°C rise. Organic activity also increases with temperature and leads to the production of higher concentrations of soil CO_2 and organic acids. These increase rates of carbonation, solution and hydrolysis. The highest rates of chemical weathering are associated with regions that have relatively high mean temperatures, and where high annual precipitation ensures the availability of water to act as a reactant and solvent. Topography affects temperature and water availability at regional and local scales, and the relationship between climate zones and weathering regimes is disrupted by mountain ranges.

The mineral composition of the parent material determines the type of chemical weathering. Minerals such as olivine and pyroxene, which are rich in iron and other metallic ions, are very susceptible to breakdown by solution, oxidation and hydrolysis. Quartz, which contains few metallic ions, breaks down much more slowly. Physical properties of the parent material, such as jointing, bedding planes and fractures, increase the area available for chemical weathering and allows water to penetrate cracks and fissures.

MASS WASTING

Mass wasting refers to the various ways in which regolith, soil and rock move down-slope under the force of gravity. It does not include transport by moving water, air or ice, although these may act as lubricants, moving with the sediment in mudflows and avalanches, or causing periodic swelling and shrinkage of slope material. Mass wasting processes operate over a huge range of spatial and temporal scales. For example, individual particles may be transported a few centimetres over hundreds of years by soil creep, while landslides involving whole mountainsides may occur within a few minutes. The largest landslide on Earth is the Saidmarreh slide in south-west Iran. This took place 10,000 years ago and involved a mass of limestone 15 km long, 5 km wide and at least 300 m thick, which travelled several kilometres (Summerfield, 1991).

Forces acting on a block of slope material

One of the most obvious controls on the down-slope movement of slope material is the steepness of the slope. The gravitational force acts vertically downwards, but the slope material cannot move vertically downwards – it must either stay where it is or move in a down-slope direction. The forces acting on a block of weathered slope material are shown in Figure 4.2. The gravitational force can be divided into two components: a down-slope component and a slope-normal component. The **down-slope component** exerts a **shear stress** on the block. This shear stress acts on the block in the same plane as the slope (parallel to the slope surface), exerting a down-slope pull on the block. The **slope-normal component** acts at right angles (normal) to the slope and resists movement by 'holding' the block in place. Another resisting force counteracting the down-slope component of gravity is frictional resistance between the block and the slope.

The relative size of the two components of the gravitational force is determined by slope angle. On gentle slopes, the slope-normal component is greater than the

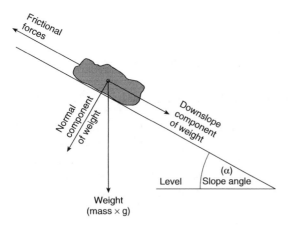

Figure 4.2 Driving and resisting forces acting on a block of slope material.

down-slope component. On steeper slopes the down-slope component of gravity dominates, increasing the shear stress on the block, and promoting movement. Whether or not the block will actually move is dependent on the balance between the down-slope component (**driving force** – promoting movement) and the combined effects of the slope-normal component and frictional resistance (**resisting force** – resisting movement).

Shear stress and shear strength

Unfortunately, predicting the stability of slopes is not simply a matter of measuring slope angle. Things are greatly complicated by interactions between the particles forming the slope material, and the moisture held in the pore spaces between particles. This means that as well as causing down-slope movement, the applied shear stress can also cause material to deform. This leads to various types of flow, slumping and slow creep. Depending on the internal resistance of the slope material, and the friction with the slope itself, the application of a shear stress can result in deformation and/or sliding. The resistance to deformation, or **shear strength**, depends on the properties of the material and the prevailing environmental conditions. Shear strength is partly determined by the amount of internal frictional resistance *within* the slope material. For example, shear strength increases when grains are tightly packed together.

This increases the contact between individual grains and therefore the friction between them. **Cohesive forces** between particles can be very significant where silt and clay-size particles are involved. The physical and chemical structure of these particles leads to the development of attractive **electrochemical forces** between them. Under some circumstances these can greatly increase the shear strength of slope materials.

There are various dynamic processes that act as triggering mechanisms for mass movements. These include earthquakes, increased slope loading due to rainfall (Brooks and Richards, 1994), and wildfire, which alters the soil hydrological properties and reduces protection of the surface by vegetation (Prosser and Williams, 1998). Other processes are also significant, such as slope loading caused by the accumulation of debris, changes in vegetation cover and slope steepening, as a result of undercutting at the base of the slope by fluvial erosion.

Moisture content is very important on soil covered slopes because water can act as a lubricant when the pore spaces between particles become saturated. The pressure of the water – termed a **positive pore water pressure** – tends to force the particles apart. This reduces internal frictional resistance and therefore shear strength. By contrast, as the material dries out, only the smallest pores are filled with water and a **negative pore water pressure** develops. The water is held in these small pore spaces by strong suction forces. A negative pore water pressure can *increase* the shear strength of the material and is most effective for the smallest clay and silt-sized particles.

Given the importance of pore water pressure, many landslides on soil-covered slopes are driven by rainstorms. Pore water pressure is highly dynamic, changing rapidly over time in response to inputs of precipitation. Although the effectiveness of a given storm is partly dependent on its intensity and duration, a very significant factor is the moisture status of the slope material prior to the storm. Where there has been a high antecedent rainfall, it may only take a small event to trigger the down-slope movement of unstable materials.

Types of mass movement

Different types of mass movement are illustrated in Figure 4.3. Some of these, such as slides, falls and certain types of flow, occur over a short period of time (minutes to days) while non-rapid mass movements include heave, creep and solifluction. The term 'land-slide' is in wide general use but actually describes several different types of mass movement.

True **slides** are characterised by a well defined **shear plane** between the moving material and the slope. The shear plane is a two-dimensional surface which lies parallel to the slope for translational slides (Figure 4.3a) but has a concave form, a little like the upper surface of a spoon, for rotational slides (Figure 4.3b). Rotational slides tend to be more characteristic of thick layers of homogeneous, cohesive materials.

Flows differ from slides in that no shear plane exists. Although the greatest shear occurs at the base of the flow, nearly all the movement is within the body of the flowing mass. The liquid content of flows varies. While the moving material in mudflows is in an almost entirely fluid state, earthflows are part solid, part liquid, and move at almost imperceptible rates. Regolith **debris avalanches** are essentially dry flows, being distinguished from slides by the inter-particle movement within the flow.

The more rapid types of flow usually have well defined boundaries. They frequently follow the line of pre-existing gullies, with a basin-shaped source area and a long, relatively narrow flow track that leads to a debris accumulation zone at the toe (Figure 4.3c). Slower earth flows create crescentic scars at the soil surface. **Falls** (Figure 4.3d) involve the vertical displacement of loose boulders, rock fragments and finer material.

Solifluction is the slowest type of flow and involves the down-slope movement of saturated

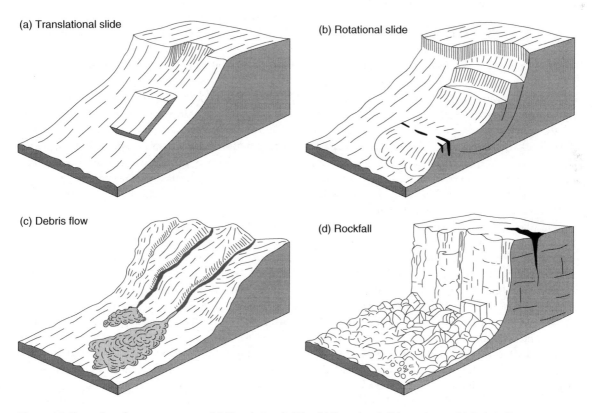

Figure 4.3 Examples of mass movement. (a) Translational slide. (b) Rotational slide or slump. (c) Debris flow. (d) Rockfall. After Huggett (2003).

slope materials. It is closely associated with **soil heave**, where slope material is vertically displaced, perhaps by the swelling and shrinking of clay particles, or the growth and melting of ice crystals. Repeated cycles lead to a net downward movement of soil material. Over longer time scales, regolith is subject to processes of **creep**, where material slowly moves down-slope under the force of gravity.

Landslides and debris flows are the dominant sediment production process in many mountain drainage basins (Benda and Dunne, 1997) and represent an important source of bedload and suspended load to river channels. Rapid mass movements provide a spatially discrete input of coarse sediment to the channel, affecting channel morphology at a localised scale. These inputs also increase the supply of coarse sediment to the channel, leading to increased rates of transport and sediment storage in the channel network (Benda and Dunne, 1997).

WATER EROSION ON HILLSLOPES

Water erosion of the soil surface is brought about by the action of falling raindrops and surface flow, which may move as a sheet across the surface or be concentrated in rills or gullies. Subsurface flow is also significant in hillslope erosion. Soil erosion provides the main source of the fine suspended sediment that is transported by river channels (clays to fine sands). Where flow is concentrated, larger material can also be transported, for example where deep gullies erode into the coarser subsoil.

There are many interrelated variables affecting rates of erosion, including climate, parent material, relief, tectonic setting, vegetation cover and human activity. For purposes of simplification, these can be considered in terms of the erosivity of the eroding agent and the erodibility of the soil surface (Figure 4.4). **Erosivity** is a measure of the capacity of an eroding agent, such as rainfall or overland flow, to erode the soil surface. It is dependent on the available kinetic energy (defined below), which is determined by factors such as rainfall intensity, raindrop size, flow depth and slope angle. **Erodibility** refers to the susceptibility of the soil surface to erosion and is dependent on the properties of the soil itself, such as soil texture (relative proportion of sand, silt and cohesive clay particles). As you will see later in this chapter,

erodibility is also dependent on the amount and type of vegetative cover and on land use practices.

Rain splash erosion

Falling raindrops, in common with all moving objects, possess **kinetic energy**. This energy is often sufficient to detach soil particles when raindrops hit the soil surface, although a large proportion of the energy is used in compacting the surface and creating an impact crater. The amount of kinetic energy a moving object has is determined by the mass of the object (this relates to the size of the raindrop) and its velocity, as shown below:

$$\text{Kinetic energy} = \frac{1}{2}mv^2$$

where m = mass and v = velocity.

You have probably heard of the terminal velocity of an object. When a raindrop starts to fall it accelerates, due to the force of gravity, from an initial velocity of zero. This gravitational acceleration takes place at a rate of 9.81 m s^{-2}, known as the gravitational constant, g. However, the raindrop does not continue to accelerate indefinitely, because it is also subject to a **drag force** as it moves through the atmosphere. This drag force acts vertically upwards, and increases with the velocity of the raindrop until it is equal and opposite to the gravitational force. This does not stop the raindrop, but it does stop it accelerating, with the result that the raindrop continues to fall at a constant, or terminal, velocity.

Raindrops do not always reach their terminal velocity and factors such as wind speed and turbulence may increase or decrease their effectiveness when they land. Also important is the presence, percentage coverage and type of vegetation. This intercepts rainfall and breaks the fall of raindrops, reducing their kinetic energy before they reach the ground surface. Rain splash erosion is therefore most effective in areas where vegetation does not entirely cover the ground surface. When raindrops fall on a sloping surface, there is a net movement of material down-slope; this increases with slope angle. Another important factor is the soil type – soils with a high silt/clay content offer the most resistance to erosion because of their cohesive nature.

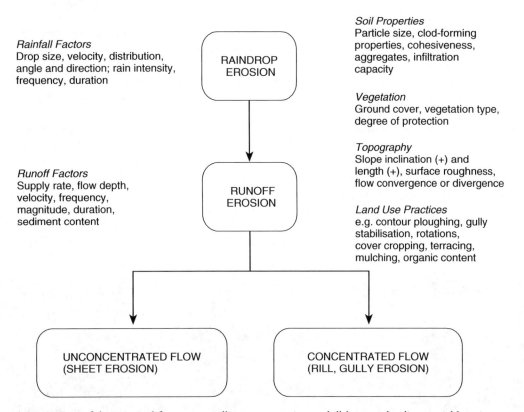

Erositivity Factors

Erodibility Factors

Rainfall Factors
Drop size, velocity, distribution,
angle and direction; rain intensity,
frequency, duration

RAINDROP
EROSION

Soil Properties
Particle size, clod-forming
properties, cohesiveness,
aggregates, infiltration
capacity

Vegetation
Ground cover, vegetation type,
degree of protection

Runoff Factors
Supply rate, flow depth,
velocity, frequency,
magnitude, duration,
sediment content

RUNOFF
EROSION

Topography
Slope inclination (+) and
length (+), surface roughness,
flow convergence or divergence

Land Use Practices
e.g. contour ploughing, gully
stabilisation, rotations,
cover cropping, terracing,
mulching, organic content

UNCONCENTRATED FLOW
(SHEET EROSION)

CONCENTRATED FLOW
(RILL, GULLY EROSION)

Figure 4.4 Summary of the principal factors controlling water erosion on hillslopes and sediment yield to river channels. After Cooke and Doornkamp (1990).

Sheetwash erosion

When significant overland flow occurs, water flows over the surface in thin layers as so-called **sheet flow**. This is a somewhat misleading term because the flow is rarely of a uniform depth, being characterised by deeper, faster threads of flow that result from micro-scale variations in the surface topography.

The down-slope flow of water exerts a shear stress on the soil surface. Erosion takes place when this shear stress is sufficient to overcome the resistance of the soil surface. Although the erosivity of the flow increases with depth and velocity, the shallow depth of overland flow and the roughness of the soil surface mean that the shear stress is not always sufficient to erode soil particles.

As a result sheetwash is only really effective on steep slopes and smooth bare soil surfaces (Morgan, 2005). However, raindrop erosion is very effective as a detachment mechanism, allowing material to be entrained (set in motion). Since the transport of soil particles requires less energy than their initial entrainment, this material can be carried by the flow until it is deposited. The combined action of raindrop erosion and sheet erosion can therefore erode a significant volume of soil from large areas of sloping land. Sheetwash tends to be dominated by fine material with a diameter of less than 1 mm (Morgan, 2005), which contributes to the suspended load of rivers. Since soil is removed in thin layers, this type of erosion may go undetected for some time.

Rills

If the flow is sufficiently concentrated, a critical shear stress may be reached at which small micro-channels called **rills** start to form. Some well developed rills, formed in a road cutting, are shown in Plate 4.1. Rills vary in size with widths of between 50 mm and 300 mm and depths of up to 30 mm (Knighton, 1998).

The critical conditions under which rills start to form can be considered in terms of a **critical shear stress** after Horton's (1945) theory of slope erosion by overland flow. It should be noted that this applies mainly to sparsely-vegetated dryland environments where intense rainfall and overland flow occur on a fairly regular basis. The diagram in Figure 4.5 represents overland flow occurring on a slope. The depth of flow increases with distance from the drainage divide, as flow accumulates in a down-slope direction (this has been exaggerated for clarity in the diagram). Since shear stress increases with depth, there must be a critical point on the slope at which the shear stress is great enough to allow incision to occur. This point is reached a **critical distance (X_c)** from the drainage divide, where the flow reaches a **critical depth (d)**. X_c varies from slope to slope according to the balance between erosivity and erodibility.

Above this point on the slope is a **belt of no erosion**. Incision can occur below this point, and parallel rills start to form in the **belt of active erosion**. Further up the slope these features tend to be discontinuous and ephemeral, being destroyed by inter-rill erosion or wall collapse. The eroded sediment is carried down-slope by the flow, reducing the energy available for further incision. If the transport capacity of the flow is exceeded, deposition starts to

Plate 4.1 Rills developed in a road cutting, South Africa.

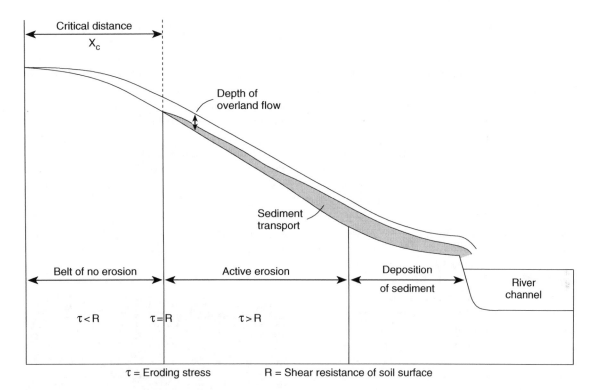

Figure 4.5 Horton's model of overland flow and rill formation. Adapted from Horton (1945).

occur in the form of small fan-shaped features. Horton called this the **zone of deposition**.

In the field, these zones are not as clearly defined as might appear from the diagram because soils are typically very heterogeneous. Even at the micro-scale, there is considerable variation from place to place in slope, roughness, infiltration capacity, cohesiveness and other factors affecting erodibility. As a result, complex spatial relationships exist between areas of erosion and areas of deposition.

Rills may develop into more permanent features under favourable conditions. They are significant in the initiation of new stream channels when network extension occurs, or where a surface has recently been exposed, perhaps as a result of glacial retreat or volcanic eruption. In order for a permanent channel to form, a sufficient concentration of flow is required. This can happen when one rill becomes dominant over neighbouring rills, incising at a faster rate, concentrating flow at depth and leading to further incision.

Even when permanent channels do not form, rill erosion, together with sheetwash and rainsplash erosion between rills (inter-rill erosion), results in the net removal of material from slopes. The concentrated flow in rills can transport larger soil particles, and even small rock fragments (Poesen, 1987). Rills may account for much of sediment removal from a hillside, although this depends on the spacing of rills and the extent of the area affected (Morgan, 2005). This can lead to the loss of soil fertility and productivity when erosion proceeds more rapidly than soil formation. New soil is produced at a rate of a few millimetres a century, whereas a single storm can result in the removal of several centimetres of soil (Woodward and Foster, 1997).

Gullying

Gullies are relatively permanent ephemeral channels. They are most commonly found in arid and semi-arid environments, where precipitation is highly seasonal and vegetation cover is sparse. Gullies are morphologically different to stream channels, being relatively deep and narrow, with steep sidewalls and a stepped channel slope. They typically range in size from depths of 0.5 m up to 25 m or 30 m (Soil Science Society of America, 1996) although there is no clearly defined upper limit, and the distinction between large gullies and some ephemeral stream channels can be somewhat vague (Poesen *et al.*, 2002). Gullies are often connected to the river system and provide an effective link between upland areas and channels, allowing the rapid transmission of water and sediment into river systems. In dryland environments gully erosion is an important sediment source, contributing an average of 50-80 per cent of the overall sediment production (Poesen *et al.*, 2002).

Smaller features, intermediate in size between rills and gullies, also exist. These are called **ephemeral gullies** and an example is shown in Plate 4.2. Ephemeral gullies are defined by the Soil Science Society of America (1996) as small channels that are eroded by concentrated overland flow and that can easily be filled by normal tillage, only to reform in the same location by additional runoff events.

Arroyos are gully-like features that are cut into debris-choked valleys. Evidence from many arid and semi-arid areas in the south-western United States suggests that these features formed rapidly after the mid-nineteenth century as a result of increasing soil erosion. Associated with gullies are **badlands**: high-relief areas that are intensively dissected by gullies and are useless for agriculture. Badlands form on unconsolidated sediments, or poorly consolidated rocks in sparsely vegetated areas and may be initiated by gully erosion. They are associated with arid and semi-arid climates but can also form in more humid climates.

The gully head

Erosion is focussed at the **gully head**, the sharp break in slope at the upslope end of the gully. In the dramatic example shown in Colour Plate 15, overland flow is occurring over a large area (the green area is totally submerged), and 'waterfalls' can be seen where the flow

plunges over multiple gully heads. Overland flow erodes the lip of the gully head as the water flows over it before falling into the plunge pool at its base, where deepening and undercutting of the headwall take place. This undermines the headwall and allows the gully head to retreat further up-slope. The steep sidewalls of the gully head are highly susceptible to various types of mass movement, especially when saturated during extreme events. Subsurface processes are also very significant in gully head retreat. Subsurface flow moving towards the gully head can weaken the walls, and the development of piping is common. The collapse of pipes further contributes to gully head retreat.

Under certain circumstances, gullies can extend rapidly upslope and tributary gullies may also form.

Plate 4.2 Ephemeral gully near Ilminster, Somerset, England. Photograph by Mike Simms.

A considerable depth of material can be removed over a short period of time, even during a single runoff event. In extreme cases, rates of retreat can be tens of metres a year (Bull and Kirkby, 2002). As a result gullying can lead to serious management problems, which are discussed in the following section.

Variables that control rates of retreat include slope steepness and the drainage area up-slope from the gully. The location of the gully head changes according to the rate of sediment supplied from up-slope and the rate at which erosion takes place at the gully head. Gullies can become partly infilled if net deposition occurs.

Gully formation

It was originally thought that gullies developed from enlarged rills. However, the process is rather more complex and various theories have been developed on the basis of observations. Gullies are thought to form when a break in the vegetation cover allows erosional hollows to form, in which water accumulates. If sufficient flow concentration occurs, an incipient gully head, or **headcut**, forms. Erosion is focused at this point and, if overland flow occurs on a regular basis, material is eroded at the headcut to deepen and enlarge the hollow. This produces a more permanent feature, further concentrating flow and leading to more erosion as a result of positive feedback (Leopold *et al.*, 1964). Once a gully channel has formed, retreat takes place by the mechanisms described above, as long as the rate of erosion exceeds the sediment supply. Gully development is also associated with landslides that leave deep steep-sided scars. These can be subsequently occupied by flowing water (Morgan, 2005).

Not all gullies develop exclusively by surface erosion. Subsurface flow is also important, and there are a number of observations of gullies that have formed as a result of pipe collapse in various soil materials and under different climatic conditions (Morgan, 2005). Harvey (1982) suggested that the occurrence of piping and tunnelling is mainly controlled by the characteristics of the soil materials. At depth, differential porosity and soil strength allow for the development of preferential flow paths and promote pipe development. The presence of deep tension and desiccation cracks allow concentrated overland flow to penetrate the soil surface.

MANAGEMENT PROBLEMS ASSOCIATED WITH ACCELERATED SOIL EROSION

Causes

Under natural vegetation, there is an approximate equilibrium between the rate at which new soil is created and the rate at which soil is eroded. Rates of soil formation vary between different environments, but are thought to average approximately half a tonne per hectare per year (Troeh *et al.*, 2003).

Land use changes such as deforestation, tillage, cropping and the conversion of land to pasture or rangeland all increase the erodibility of the soil. **Accelerated soil erosion** occurs when rates of soil erosion increase as a direct or indirect result of human activity. This is a major environmental problem worldwide. Natural vegetation breaks the fall of raindrops and helps to bind loose soil on slopes. It also reduces the frequency of overland flow by encouraging infiltration. Organic material, in the form of dead and decaying plant material, plays several important roles: it holds the topsoil together, increases permeability and provides a supply of nutrients. The removal of natural vegetation greatly reduces soil protection, and rates of soil erosion may accelerate manyfold. Soil losses from land covered by perennial vegetation are a fraction of a tonne per hectare per year. However, losses from bare cultivated fields in excess of 450 tonnes per hectare have been reported.

Some environments have a greater degree of sensitivity to land use change than others. Sensitivity is determined by a number of variables including slope steepness, slope length, the seasonality and intensity of rainfall and soil characteristics. The deforestation of steep slopes in tropical and mountain environments such as Nepal, the Loess Plateau of central China, the Ethiopian Highlands and Madagascar has had devastating consequences. In these and many other parts of the world problems are worsened by rapid population growth, agricultural expansion, more intensive agricultural production, and a growing demand for timber and fuel wood.

Even in the UK, cultivation of winter crops has led to sheet and rill erosion on sloping arable land in many locations (Plate 4.2). Winter cropping means that arable land lies bare and exposed to more intense winter precipitation (Boardman *et al.*, 1990).

Impacts

There are many environmental impacts associated with accelerated soil erosion. **On-site impacts** refer to the effects seen at the site of the erosion, while **off-site impacts** result from the huge increases in sediment supplied to river channels. Off-site impacts are considered more fully in Chapter 5.

Once the fertile topsoil has been removed, the lower soil layers are exposed. These have a poor structure and are low in organic matter and nutrients. As a result they are less permeable, which increases overland flow and leads to yet more erosion. In this way several centimetres depth of soil can be removed by a single rainfall event (Woodward and Foster, 1997). Gullying can remove vast quantities of soil in a short period of time. The landscape becomes deeply dissected and may no longer be suitable for tillage or grazing. Gullying can also destroy bridges and cultivation terraces.

Plate 4.3 shows numerous gullies in eastern Madagascar, known locally as *lavaka*. Madagascar has experienced widespread deforestation since colonialisation in the 1900s. It is now estimated that over 90 per cent of the original forest cover has disappeared as a result of clear-cutting, burning, farming, logging and clearing of land for settlements (Bakoariniaina *et al.*, 2006).

Plate 4.3 Severe gullying in eastern Madagascar. Photograph by Rhett Butler, WildMadagascar.org.

Soil erosion is so severe that rivers run bright red with the soil that is washed into them after heavy rain, a sight that has been described by NASA astronauts as Madagascar 'bleeding into the ocean'. The problem is greatly exacerbated by frequent intense rainfall associated with tropical storms. Although deforestation in Madagascar may partly be due to changes in global and local climate, the vast majority is a result of human activities.

The exceptionally high sediment loads carried by Madagascar's rivers have resulted in vast volumes of sediment building up in lakes and river channels. The estuary of the largest river, the Betsiboka, has become so badly silted that oceangoing ships can no longer navigate it and must now berth at the coast. Silt also builds up at the base of hillslopes, covering the surface of rice cultivation areas. Deforestation continues on a massive scale, threatening one of the most biologically rich areas on Earth.

Management strategies

Rates of soil erosion can be reduced by land management strategies. Terracing has been practised for thousands of years in many parts of the world. Low walls are built from stone or other materials and run parallel to the contours. Soil is placed behind these walls to form a series of benches for cultivation. Contour ploughing, with plough furrows running across the slope rather than up and down it, also helps to reduce soil loss.

Protection can be provided for the bare soil in between cultivation rows by the planting of cover crops. The practice of crop rotation between growing seasons involves the periodic cultivation of lower yielding crops that provide greater soil protection. This allows some recovery from the effects of more intensive cultivation in the intervening years.

Gullying, where this has developed or increased, can be very difficult to control. Gullies can be blocked by building low walls across them to impede flow and trap sediment. If given the chance to re-vegetate, gullies can be stabilised, although this process can take a number of years.

Although soil erosion can be reduced using these strategies, their implementation may be problematic in

areas where a high population density imposes severe demands on the land. Soil erosion is a complex problem that is not just governed by physical processes; social, economic and political factors are also of great significance.

MONITORING RATES OF EROSION

A number of different techniques are used in the assessment of soil erosion. These include the collection of data in the field, laboratory simulations of artificial slope environments and numerically based soil erosion models. It is important to note that all these measurements are subject to error. This affects the degree of confidence that can be placed in theories and models that are developed on the basis of the measured data. Although the degree of error cannot be quantified, it is possible to determine the amount of variability by repeating the same measurements a number of times. This allows confidence limits to be applied to the data.

Field measurements

Field measurements are carried out at different scales and usually involve measuring the amount of soil that is eroded from a known area. For example, soil losses can be monitored by collecting the runoff and sediment generated from the area of a test plot. The size of individual plots depends on the nature of the investigation, a standard size being 22 m long by 1.8 m wide (Morgan, 2005). Plot boundaries are usually isolated from the surrounding area so that only sediment generated from the plot itself is included in measurements. Water and sediment flow into a trough and collecting tank at the down-slope end of each plot. The volume of sediment eroded from the area over a given period of time can then be assessed and related to runoff volume. Rainfall is also monitored using rain gauges, although a rainfall simulator may be used to control rainfall inputs to the plot area. These basically consist of one or more nozzles that are mounted a given height above the soil surface on a scaffolding framework. Rainfall characteristics, such as drop size, velocity and intensity can then be simulated, although no rainfall simulator accurately simulates all the characteristics of natural rainfall (Hall, 1970). For example, the height is not sufficient in most cases to allow simulated raindrops to reach their terminal velocity, resulting in reduced kinetic energy. Although this can be compensated for by feeding water to the nozzles under increased pressure, this produces unrealistic drop size distributions (Morgan, 2005). There are also a number of problems associated with measurements from test plots, such as silting of the collection tank and connecting pipes, overflowing of tanks, and the development of rills along preferential flow paths at the edges of plots (Morgan, 2005).

The use of plots allows experiments to be carried out to investigate the effects of one or more variables on rates of soil erosion. For example, a comparison could be made using plots with different types or amounts of vegetation cover. There are several potential sources of error in these measurements. Rates of soil erosion are also highly variable over a small area. For these reasons it is important to use at least two identical test plots for each experiment and calculate an average soil loss value.

Another method for estimating soil loss is to install erosion pins. These are nails 250 mm to 300 mm long that are set into the ground. The length that protrudes above the soil surface then provides a point of reference for assessing rates of soil loss over time. Problems include the fact that pins are easily destroyed by livestock and wildlife, or removed by people who have an alternative use for the nails. The relocation of pins can also be difficult during subsequent experiments.

Where rills and gullies are present, changes in their 'volume' can be monitored. Measurements of the cross-sectional area (width × depth) made at regular intervals along the length of a rill or gully are combined with length measurements to calculate the volume of soil that has been removed. Once this is known, rates of soil loss over a wider area can be estimated according to the number of rills or gullies found there. Volumetric estimates can also be made from aerial photographs, using photogrammetric techniques, or digital elevation models. Laser-based techniques can also be used to provide high-resolution data.

Similar techniques can be used to assess the volume of sediment removed by various types of mass movement process. The frequency of occurrence on a given slope can be estimated by deriving an approximate age for former landslide scars. Measurements are subject to error, since it can be difficult to define the edge of features. In the case of rill erosion, estimates from these measurements ignore the contribution from inter-rill erosion and can lead to underestimates of 10–30 per cent (Morgan, 2005).

Rates of erosion can also be inferred from measurements of suspended sediment concentration and flow discharge made at the drainage basin outlet. Alternatively, lake or reservoir surveys can be made to estimate the volume of sediment that has been deposited over a given period of time. These techniques are discussed in Chapter 5.

Soil erosion models

A number of numerical models have been developed to simulate soil erosion. The most widely used model is the Universal Soil Loss Equation (USLE). This is an empirical model, which means that the relationships between the various model components and rates of erosion have been derived from statistical analysis of field data. It has the form:

$$A = R \times K \times L \times S \times C \times P$$

where A is the mean annual soil loss, R is a rainfall erosivity factor, L a slope length factor, S a slope steepness factor, C a crop management factor, P an erosion control practice factor and K the soil erodibility index.

The model has been very successful, due to its simplicity and to the extensive database that was used to derive it (Nearing *et al.*, 1994). USLE was first developed during the 1950s and 1960s using data from runoff plots at forty-nine research stations across the United States. When multiplied by the number of plots at each station, and the number of years of data, this resulted in over 10,000 plot years of data. Data from other areas, together with additional runoff plots and rainfall simulator studies, have also been incorporated into subsequent versions. Disadvantages of this approach are that it only calculates the long-term soil loss over a period of time and that spatial variations in soil loss are not simulated.

More sophisticated physically-based models include the US Water Erosion Prediction Project (WEPP) and European Soil Erosion Model, EUROSEM. These models are spatially distributed, representing the slope as a number of segments, between which transfers of water and sediment are simulated. These transfers are calculated at a certain time step, representing a period of time such as a day. For example, if rainfall occurs over a simulated slope on a given day, runoff is computed. If there is sufficient runoff, soil detachment, down-slope transport and deposition are simulated. Changes in the soil volume and biomass of each segment are also calculated.

CHAPTER SUMMARY

Sediment is produced in the headwater regions of the source zone by processes of weathering, mass movement and erosion. Weathering involves the physical break-down of rocks at the Earth's surface and produces material called regolith. This is transported down-slope, under the force of gravity, by processes of mass wasting. These include rapid mass movements, such as slides and debris flows, together with the much slower processes of creep and solifluction. Sediment is also produced by the erosive action of water, ice and wind. Processes of water erosion include rain splash, sheetwash, rilling and gully-ing. Soil erosion is a natural process, but it can be accelerated by human activity, with rates of soil removal exceeding rates of soil formation. Accelerated soil erosion is a major environmental problem worldwide. In order to assess rates of soil loss, various monitoring techniques are used. Models have also been developed to simulate erosion and soil loss.

FURTHER READING

Introductory texts

Kirkby, M.J., 2005. Hillslope processes and landscape evolution. In: J. Holden (ed.), *An Introduction to Physical Geography and the Environment*. Pearson Education., Harlow, pp. 249–77. Covers weathering and mass movement in slightly more detail than is provided here.

Summerfield, M.A., 1990. *Global Geomorphology: An Introduction to the Study of Landforms*. Longman, Harlow. Good, clearly written chapters on weathering and slope processes.

More advanced texts

Morgan, R., 2005. *Soil Erosion and Conservation*. Blackwell, Oxford. Comprehensive coverage of soil erosion processes, monitoring and modelling, and management.

Poesen, J. *et al.*, 2002. Gully erosion in dryland environments. In: L.J. Bull and M.J. Kirkby (eds), *Dryland Rivers: Hydrology and Geomorphology of Semi-arid Channels*. John Wiley & Sons., Chichester, pp. 229–62. Provides a summary of current understanding of gully erosion in dryland environments.

Web sites

The Soil Erosion Site, www.soilerosion.net. Here you will find a wide range of material relating to all aspects of soil erosion. With contributions from nearly 50 soil erosion experts, the site includes background information, links to photos and videos, and more specialist material.

5

LARGE-SCALE SEDIMENT TRANSFER

Sediment generated by primary erosion is transferred downslope and onwards through the channel network. However, only a small proportion of this sediment actually exits the drainage basin. This is because a significant volume of sediment enters storage when it is deposited along the way. Sediment can be released from storage, when it is re-eroded at a later stage, and an individual particle of sediment can be stored and remobilised many times as it is transported through the fluvial system. In this chapter you will learn:

- How sediment moves through fluvial systems.
- Where sediment is stored, and why this is important.
- Controls on the amount of sediment that is delivered to the oceans.
- Sediment budgets.
- Human impacts on the sediment system.

SEDIMENT TRANSFER

Sediment transfer from hillslopes to channels

The term **primary erosion** is used to describe the initial, or *in situ* erosion of rock, regolith and soil. It does not include the re-erosion of material that has been deposited, for example, at the base of a hillslope. Sediment that has

been transported downslope and deposited on or at the base of slopes is called **colluvium**.

There is an important linkage between the erosion of sediment from hillslopes and its transfer to channels and valley floors. The effectiveness of this transfer is dependent on the degree of **hillslope–channel coupling**. In a coupled system there is a direct transfer of sediment from slopes to channels. This is typically the case in headwater regions, where narrow valleys are bordered by steep hillslopes. Coupling also occurs when a channel erodes the valley margin.

Sediment transfer to channels is much more limited in decoupled systems. Further downstream, where valleys widen and channels are bordered by floodplains, sediment is stored at the base of hillslopes or on floodplain surfaces. In decoupled systems, colluvial sediment makes only a very small contribution to the river's sediment load.

The degree of coupling varies according to sediment size. Finer sediment is more mobile and can be transported over greater distances. For a given slope–channel system, the degree of slope–channel coupling is often stronger for fine sediment than it is for coarse material.

Modes of sediment transport in river channels

The sediment supplied to most river channels varies greatly in size, from microscopic clay particles to

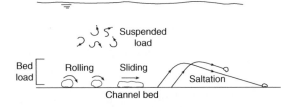

Figure 5.1 Modes of sediment transport.

large boulders. As you may be aware, there are different mechanisms involved in the transport of coarse and fine sediment within river channels. These are illustrated in Figure 5.1 and have important implications for the way in which sediment of different sizes is transferred through the system.

Coarse material – typically coarse sands, fine gravels and larger particles – is moved along the bed of the channel as **bedload**. Particles are in continuous or regular contact with the channel bed and move by rolling, sliding or in a series of 'hopping' motions called saltation.

The **suspended load** consists of finer material and usually includes clays, silts and sands. This material is carried aloft, suspended above the channel bed by turbulent eddies, and is transported downstream in the main body of the flow. The finest fraction of the suspended load is called the **wash load** and consists of tiny clay particles with diameters of less than 0.0063 mm. This material is so fine that it can remain in suspension at very low flow velocities, even when water movement is barely perceptible. Wash load sediment can be carried over many kilometres in a matter of hours.

Material is also transported in solution as the **dissolved load**. These solutes are derived from a number of sources, including rock and soil weathering, the atmosphere, biosphere and human activity.

Most of the research discussed in this chapter has focused on the transfer and delivery of suspended sediment (including the wash load). One of the main reasons for this is that bedload transport is notoriously difficult to monitor or calculate accurately. As a result, bedload has rarely been included in estimates of sediment yields exiting river basins. While the material that is discharged to the oceans is predominantly composed of fine sediment, it should not be assumed that this is always the case. The transfer of coarse sediment is discussed on pp. 60–61.

SEDIMENT YIELD

The **sediment yield** is the total amount of sediment that exits a drainage basin over a given period of time, usually a year. It is commonly expressed in units of tonnes per year ($t\ y^{-1}$). (Box 5.1 explains how the sediment yield of a river can be estimated.) Huge sediment yields are associated with the largest rivers. The lower Amazon, which drains an area of 6.1 million km^2, has a sediment yield of 1,200 million $t\ y^{-1}$ (Meade *et al.*, 1985). This amount of sediment would occupy a volume of approximately 0.43 km^3.

To compare rivers draining different areas the **specific sediment yield** is used. This is the sediment yield per unit area of the drainage basin (usually per square kilometre) and is calculated from:

Specific sediment yield ($t\ km^{-2}\ y^{-1}$)   =
Sediment yield ($t\ y^{-1}$) / Drainage basin area (km^2)

The specific sediment yield of the Amazon is therefore just under 197 $t\ km^{-2}\ y^{-1}$.

The difference between sediment yield and specific sediment yield is illustrated for two rivers in Figure 5.2. The Santa Clara River (California, United States) and the Hokitika River (South Island, New Zealand) both have a sediment yield of approximately 6 million $t\ y^{-1}$. However, the Santa Clara drains an area twelve times the size of the Hokitika basin. Clearly the Hokitika is producing a much greater volume of sediment per unit area. The corresponding specific sediment yields are 1,400 $t\ km^{-2}\ y^{-1}$ for the Santa Clara and a massive 17,000 $t\ km^{-2}\ y^{-1}$ for the Hokitika, which has one of the highest specific sediment yields in the world.

Sediment storage and the sediment delivery ratio

Only a certain percentage of the sediment produced by primary erosion actually reaches the drainage basin outlet. This is because a significant volume enters storage

Box 5.1

ESTIMATING THE SUSPENDED SEDIMENT LOAD OF A RIVER

The amount of sediment exiting a drainage basin can be determined from measurements of **discharge** and **sediment concentration**.

Establishing the relationship between discharge and sediment concentration

At the exit point, discharge is monitored on a continuous basis at a gauging station (see Box 3.1). The mean daily discharge is calculated for each day on record.

Water samples are taken at different times to determine the suspended sediment concentration, at various discharges, over a period of time. This is done using a **suspended sediment sampler** such as the DH59 sampler shown in Figure 1. A **depth-integrated sample** is taken by lowering the sampler vertically through the flow to the near bed region and back again. Water and sediment enter the intake nozzle and are collected in a sample bottle.

The sample is then filtered, dried and weighed to give a dry weight of sediment in milligrams (mg). When divided by the volume of the sample in litres (l) this gives a concentration in milligrams per litre (mg l^{-1}). Over time, a dataset of sediment concentration measurements is built up, each corresponding to a different discharge. From this, the relationship between sediment concentration and discharge can be determined using regression analysis. This relationship is of the form:

$$C = aQ^b$$

where C is sediment concentration, Q is discharge, a and b are coefficients that describe the unique relationship between discharge and sediment concentration for each river cross-section.

Using this relationship to estimate the annual sediment load

Using the equation above, the sediment concentration (C) corresponding to the mean discharge (Q) for each

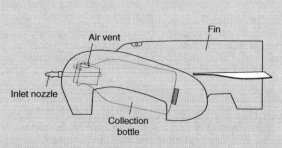

Figure 1 The DH59 suspended sediment sampler.

day can be calculated. This gives an *instantaneous* sediment concentration. To calculate the *total sediment load* passing the recording station for that day, it is necessary to multiply the instantaneous concentration by the total volume of flow for the day:

Sediment load (kg) = sediment concentration
(mg l^{-1}) × mean discharge (m^3 s^{-1}) × No of seconds
in a day

Since there are 86,400 seconds in a day, the numbers can get rather large. Therefore the daily sediment load is usually expressed in kilograms rather than milligrams (there are 1 million mg in 1 kg).

To estimate the **annual sediment load**, the process is repeated for each day of the year. The accuracy of these estimates is very much dependent on the range of flows that were sampled in deriving the relationship between discharge and sediment concentration. The sediment load will be underestimated if flood flows are not included in the data used in the regression analysis.

It is possible to record sediment concentration on a continuous basis by monitoring turbidity. Sediment concentration is not measured directly, instead **optical sensors** can be used to record how much light passes through the flow. These measurements can then be related to the concentration of sediment, since more light is absorbed at higher concentrations.

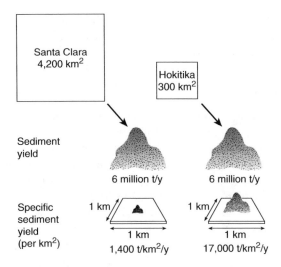

Figure 5.2 Sediment yield and specific sediment yield for the Hokitika River, New Zealand, and the Santa Clara River in California, United States. Based on data collated by Milliman and Syvitski (1992).

when it is deposited along the way. Some examples of the locations in which sediment is stored are illustrated schematically in Figure 5.3. These include hillslopes, alluvial fans, river channels, floodplains, deltas and lake bed deposits.

The proportion of eroded sediment that exits the drainage basin as part of the river's load is called the **sediment delivery ratio (SDR)**. This is the ratio of the primary erosion rate on hillslopes to the sediment yield at the basin outlet. An example is provided by research carried out by Foster *et al.* (1996) on the River Start in south Devon, England, which drains an area of 10.8 km. Upstream rates of soil erosion have been estimated at 107 t km^{-2} y^{-1}. However, the annual specific sediment yield monitored at the basin outlet is just 29 t km^{-2} y^{-1}. The difference is due to the fact that much of the eroded sediment is deposited at field boundaries and on floodplains in the lower valley. As a result only 27 per cent of the sediment produced in the headwaters actually reaches the sea – this 27 per cent is the SDR.

The River Start, which drains a small basin, has a relatively high SDR. A greater proportion of sediment is stored in larger drainage basins, and SDR values are often as low as 5 per cent (Walling and Webb, 1983). This means that only a fraction of the sediment produced by primary erosion actually exits the basin.

Some of the sediment that enters storage is remobilised by erosion at a later stage. Storage residence times vary enormously, ranging from a few seconds to millions of years or more. The length of time a sediment particle stays in a given store depends on several things. These include the geomorphic setting, the frequency of

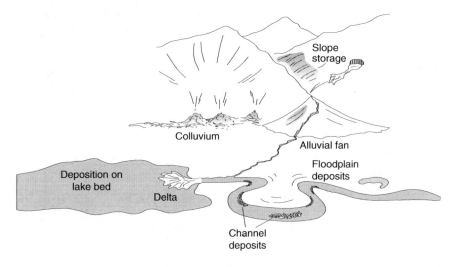

Figure 5.3 Sediment stores.

erosive events, and particle characteristics such as grain size. Sediment can be stored and re-eroded many times as it moves through the fluvial landscape.

Controls on sediment yield

Global variations in specific suspended sediment yields are shown in Figure 5.4. The data used to produce this map have come from a number of different sources, some of which are more reliable than others. It has also been necessary to extrapolate existing data to cover drainage basins for which there is no data.

Specific sediment yield is controlled by a combination of different variables, which include climate, basin area, topography, soils and geology, and human activity.

These are considered below, although it is important to note that it can be very difficult to isolate the effect of a particular variable.

Climate

In Chapter 4 you saw how rates of soil erosion are influenced by the erosivity of rainfall and runoff on slopes. With this in mind, a number of researchers have attempted to relate sediment yield to climatic variables such as mean annual precipitation or runoff. The work of Langbein and Schumm (1958) provides one example. They examined the sediment yield for several small basins in the western United States. Sediment yields, obtained from river data and reservoir surveys, were

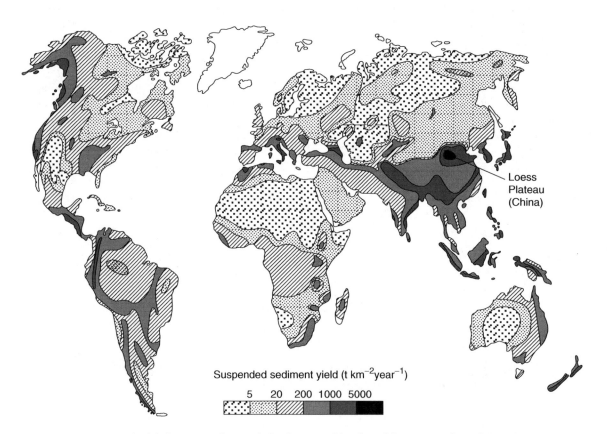

Loess Plateau (China)

Suspended sediment yield (t km^{-2}year^{-1})

5   20   200 1000 5000

Figure 5.4 Global patterns of suspended sediment yield. Adapted from Lvovitch *et al.* (1991).

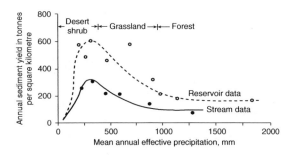

Figure 5.5 Climatic variation of sediment yield determined from gauging stations and reservoir surveys. Adapted from Langbein and Schumm (1958).

related to the annual effective precipitation for each basin (calculation of an effective precipitation incorporates losses due to evaporation,[1] which vary from basin to basin).

The relationship is shown in Figure 5.5. Maximum sediment yields were found to occur where the annual effective precipitation was about 300 mm, which corresponds to a semi-arid climate. The authors suggest that, in these environments, the vegetation cover is too sparse to provide adequate protection from intense rainfall. For rivers in dryer climates, sediment yields are low because rainfall is insufficient to cause significant erosion. In wetter areas, where the annual precipitation exceeds 300 mm, a more extensive cover of protective vegetation becomes established, reducing sediment yields.

This relationship is not universal. Other researchers, working in different parts of the world, have observed a variety of responses in sediment yield to changing precipitation. A review of this research was carried out by Walling and Webb (1983), who concluded that current evidence emphasises that there is no simple relationship between climatic variables and sediment yield. More recent research indicates that climate may be of secondary importance to the influence of drainage basin size and topography.

Drainage basin size and topography

The size of a drainage basin determines the space, or volume, available for storing sediment – larger drainage basins are able to store more sediment. Of course, primary erosion also generates higher sediment yields in large drainage basins. However, the *specific* sediment yield tends to decrease with increasing basin size. This means that in larger basins a greater *proportion* of the eroded sediment ends up in storage.

The relationship between basin area and specific sediment yield was analysed by Milliman and Syvitski (1992) using data for 280 rivers. These were classified into seven different groups on the basis of the maximum elevation in each basin. The graphs in Figure 5.6 clearly show the inverse relationship that exists between basin area and sediment yield within each group.

Also evident is the influence of maximum drainage basin elevation. The composite graph (lower right of Figure 5.6) shows the relative sediment yields for mountainous, upland, lowland and coastal plain rivers.

Some of the greatest specific sediment yields are observed for tectonically active regions. The rivers in group A rise in high mountains such as the Andes and Himalayas at altitudes exceeding 3,000 m. These mountain belts are all experiencing rapid rates of tectonic uplift. Associated with this are steep, sparsely vegetated slopes, highly fractured rocks and seismic activity. Erosion rates are further enhanced by highly erosive rainfall resulting from orographic uplift and there may also be significant contributions from glacial erosion. A satellite image of the Ganges–Brahmaputra river system is shown in Colour Plate 12. Both rivers rise in the Himalayas, where enormous volumes of sediment are produced. The brown area surrounding the Mouths of the Ganges is suspended sediment being discharged to the ocean. However, this is only a fraction of the sediment that enters storage on the extensive floodplains and Ganges Delta.

The rivers in group B drain the high-standing Pacific islands between Australia and Asia. This is another tectonically active region. Estimates based on data from New Guinea, the Philippines, Java, New Zealand and Taiwan suggest that *average* yields for this region may be close to 3,000 t km^{-2} y^{-1} (Milliman and Syvitski, 1992). This group includes the small, mountainous Hokitika River referred to in the Figure 5.2 example.

Glacially fed rivers, such as the rivers of western Canada and Alaska (group C) are supplied with vast quantities of sediment. Many rivers in this region are still reworking thick deposits of glacial material laid down during the last glaciation.

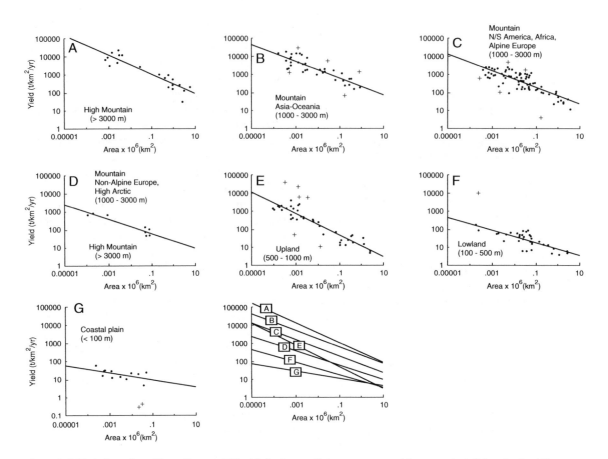

Figure 5.6 Variation of specific sediment yield with basin area for seven topographic categories of river basin. After Milliman and Syvitski (1992).

Much lower yields are associated with lowland (group F) and coastal plain (group G) rivers. Bear in mind that the scale on the sediment yield axis is logarithmic, so the differences between the different groups are greater than they might appear from the graphs. For example, the lowland St Lawrence River in Canada, which drains an area of 1.1 million km², has a yield of 4 million t y⁻¹ and a specific yield of just 4 t km⁻² y⁻¹. Specific yields of less than 0.5 t km⁻² y⁻¹ are reported for some Finnish rivers.

It is small mountainous drainage basins that have the highest specific sediment yields. Steep slopes and narrow valleys present relatively few opportunities for sediment storage. As the size of a drainage basin increases, there is a decrease in the proportion of its total area that is mountainous. This is true even for rivers that rise in the highest ranges, such as the Ganges, Brahmaputra and Amazon. It is estimated that more than 80 per cent of the Amazon's sediment load is derived in the Andes mountains (Meade *et al.*, 1985). However, only 10 per cent of the total drainage area is mountainous. Further downstream the main river and its tributaries are bordered by vast floodplains. These cover an extensive area and store unimaginable volumes of sediment.

An important implication of Milliman and Syvitski's research is that previous estimates of sediment flux to the oceans may have greatly underestimated the contribution of small, mountainous drainage basins.

Erodibility

Rates of primary erosion on hillslopes are determined by the erosivity of rainfall and runoff and the erodibility of the surface. This is influenced by factors, including geology, soil type, vegetation and human activity. Slope angle and length also affect erodibility, again highlighting the important influence of topography.

Exceptionally high rates of erosion are observed in the Huanghe (Yellow) River basin, which drains the Loess Plateau of central China (Figure 5.4). This is formed of thick Quaternary deposits of fine grained sediment called loess, which is referred to as 'perhaps the most easily erodible material available to moving water' (Milliman and Meade, 1983: 7). In combination with the effects of cultivation and heavy monsoon rainfall, this leads to very high rates of soil erosion in the region.

At the other extreme, sediment yields from continental interiors can be very low. The yields for northern Europe, Siberia and the continental interior of Canada are often less than 20 t km^{-2} y^{-1} (Woodward and Foster, 1997). These ancient shield terrains form extensive areas of low relief, resistant bedrock, and have well established vegetation cover. Not surprisingly, yields are very low for desert regions such as the Sahara and Australian deserts.

Antecedent environmental conditions

Past environmental changes can greatly alter the availability and distribution of sediment within drainage basins. This means that the linear decrease in specific sediment yield with area, seen in the graphs of Figure 5.6, is not always observed. For example, a rather different relationship exists for the rivers of British Columbia in Canada. Church and Slaymaker (1989) observed that for basins of between 40 km^2 and 10,000 km^2 the specific sediment yield *increases* with basin area. Then, as basin size increases beyond this range, there is a decrease in the specific sediment yield with increasing area. This non-linear variation can be explained by the fact that these rivers are still responding to the effects of the last glaciation, during which vast quantities of fluvio-glacial sediment were deposited in river valleys and on extensive outwash plains. Although these deposits are relatively thin in the mountainous headwaters, large

accumulations of sediment are found further downstream. Even though this material was deposited over 10,000 years ago, it is still being remobilised by the present-day rivers. Large quantities of sediment are still being released into river channels through the erosion of river banks and the immediate valley sides. In fact, this remobilised material contributes a much greater proportion of sediment to the overall yield than does the primary erosion of the land surface (Church and Slaymaker, 1989). Remobilisation of the fluvio-glacial deposits accounts for the downstream increase in specific sediment yield. As basin area increases beyond 10,000 km^2, however, the increased potential for sediment storage results in a subsequent decrease in specific sediment yield. The time scale for the ultimate dispersal of the glacial material is estimated to be tens of thousands of years (Church and Slaymaker, 1989).

A similar relationship was observed in the Yellow River basin by Jiongxin and Yunxia (2005). In this case specific sediment yield increases with basin area, reaching a maximum at about 2,000 km^2, with a subsequent decline for further increases in area. The middle part of the Yellow River basin crosses the Loess Plateau, which formed during the Pleistocene from wind-blown deposits, resulting from a strengthened East Asian monsoon. These deposits originally had a thickness of approximately 200 m, representing an enormous quantity of stored sediment. The present thickness is estimated to be 100 m and it would take approximately 22,000 years for the supply of sediment to be exhausted (Jiongxin and Yunxia, 2005).

Human activity

Human activity has greatly altered the yield of many rivers worldwide. Increases in yield are caused by accelerated soil erosion, mining and urbanisation.

The yield of an increasing number of rivers has been reduced by the construction of large dams. Sediment is deposited in reservoirs, becoming trapped behind dams. The Colorado River used to have an annual sediment yield of 120 million t y^{-1}. However, the construction of major dams on the river has led to a dramatic decrease and the annual sediment yield is now just 0.1 million t y^{-1} (Milliman and Meade, 1983). The sediment yields of an increasing number of rivers are being dramatically reduced in this way. Other examples include the Nile, Indus and Ganges. Human impacts on sediment systems are discussed further on pp. 65–67.

Time

When considering measurements of suspended sediment yields, time is an important consideration. Measurements of sediment yield can only provide a 'snapshot' of a system that operates and responds to change over a whole range of timescales. Even where several years of data exist, the sampling period is essentially instantaneous in comparison with long-term changes that might be taking place within the basin.

Extreme events, such as floods and volcanic eruptions can lead to dramatic increases in sediment yields over a relatively short period of time. For example, the average yield of the Santa Clara River is approximately 6 million t y^{-1} (one of the examples used in Figure 5.2). However, it was estimated to be 50 million t y^{-1} in 1969, the year of a major flood (Curtis *et al.*, 1973).

Over periods spanning decades or centuries, the occurrence of a few major floods can have a major influence on overall rates of sediment transfer. Indeed, episodic high-magnitude events such as landslides and floods might be a dominant sediment source (Dietrich and Dunne, 1978). The problem is that such events may not occur during the monitoring period for a given river, even where several years of data exist. This would result in the sediment yield being underestimated for that river. As you can imagine, even when a major flood does occur there are many practical problems associated with monitoring sediment loads.

COARSE SEDIMENT TRANSFER AND YIELD

Compared with suspended sediment transfer, the movement of coarse sediment through the channel network tends to be less rapid and more localised. These differences reflect the processes by which bedload and suspended load are transported by channel flow.

The energy required to transport a particle of sediment increases with particle size. Very little energy is needed to transport the fine clay particles that are washed into headwater channels during rainfall events, although the initial erosion of these fine particles does require rather more energy than is needed to transport them. In fact, even when a large amount of fine sediment has been flushed into the channel, the available energy for transporting this material is frequently in excess of what is needed. This means that rates of suspended sediment transport tend to be **supply limited**.

In contrast, it is usually energy availability that is the limiting factor for rates of bedload transport, which is **transport limited**. Depending on the size of the channel, coarser gravels and cobbles may only be moved during times of high flows, while boulders might just be moved during exceptional floods. Once set in motion, a bedload particle is usually only carried a short distance before being deposited again. There are various reasons for this, one being that the shear stress exerted at the bed of the channel varies considerably over short distances, frequently dropping below the critical value needed to transport a particular particle. Bedload movement is also interrupted when particles become trapped behind other particles at the bed of the channel.

Because of these differences, suspended sediment can be carried over much greater distances in a given period of time. This is illustrated by a study of sediment transfer that was carried out by Bogen (1995) on a glacially fed river in Norway. Observed travel times were very different for suspended sediment, which only took a day to exit the basin, and bedload, which took several decades to travel the length of the basin. Bogen suggested that valley morphology was an important control on bedload movement, especially features such as steps and over-deepened basins. These can act as coarse sediment traps, or cause a reduction in the channel gradient, and therefore the energy available for transporting coarse sediment. There may even be some reaches of channel through which bedload cannot be transported at all, unless it is broken down into finer particles.

Important here is the concept of **connectivity**. This is defined by Hooke (2003) as 'the transfer of sediment from one zone or location to another, and the potential for a specific particle to move through the system'. A high degree of connectivity exists for suspended sediment (sand and finer) and it has often been assumed that the same is true for bedload. However, there is increasing evidence to suggest that sources and transfer of bedload may in fact be much more localised. Inputs of bedload to a given channel reach include material transported from upstream, bank erosion, tributary inflows and, where slope–channel coupling exists, adjacent hillslopes. This material may only be transported a short distance before it is deposited on a channel bar, perhaps entering storage for a period of years or decades. Further transfer downstream is very much dependent on the degree of connectivity that exists between channel reaches.

Bedload material is often transported downstream in the form of **slugs**, or waves, of sediment. These can be visualised as being similar to the downstream passage of a flood wave through a reach of channel, with an increase and subsequent decrease in the volume of sediment supplied from upstream. The passage of a sediment slug through a channel reach leads to variations in rates of sediment transport and storage volumes over time. Sediment slugs exist at a variety of temporal and spatial scales. The can range from small-scale perturbations in sediment transport generated by localised bank erosion to major changes in sediment supply brought about by tectonic activity, glaciation or human influences (Nicholas *et al.*, 1995).

A common characteristic of many wandering and braided gravel-bed rivers is alternating **sedimentation zones** and **transport reaches**. These were first defined by Church and Jones (1982), who worked on the lower Bella Coola River in Canada. Church and Jones noted the occurrence of relatively unstable reaches in which there were large accumulations of sediment, stored in the form of multiple bars. This sediment entered the channel through the erosion of Neoglacial moraines during the nineteenth century, which were located several kilometres further upstream. Rather than being dispersed through the system, the material moved as an isolated wave, travelling from one sedimentation zone to another. A rapid decrease in the sediment load was caused by a decline in the supply from the glacial moraines. In addition, the transfer of sediment from upstream was blocked by the growth of a tributary fan across the main channel[2] (Church, 1983). This resulted in a net removal of sediment from the sedimentation zones further downstream. As a result, dramatic changes

in bar form and channel stability were observed in these downstream reaches. Church produced a series of maps showing how the instability gradually moved downstream between 1893 and 1974 (see Church, 1983: figure 5, p. 175).

The formation and configuration of sedimentation zones are controlled at different scales, being influenced by landscape characteristics as well as by local channel conditions. These controls include hill spurs, the influence of tributary junctions, bedrock outcrops and variations in bank resistance (Xu, 1997).

SEDIMENT BUDGETS

Sediment budgets are used to describe the inputs, transfer, storage and outputs of sediment from a geomorphological system. This could be a reach of channel, a section of hillslope or an entire drainage basin.

What is a sediment budget?

Figure 5.7 shows a conceptual representation of a sediment budget for a reach of channel. Inputs of sediment come from upstream and from hillslope–channel transfer. Further inputs are provided by tributaries joining the main channel. Sediment is output at the downstream end of the reach.

As sediment is transported through the reach some of it will enter storage, in the form of channel deposits and overbank deposits of fine sediment laid down by floodwaters. Material deposited on the inside of migrating meander bends is also stored in the form of lateral accretion deposits, which eventually become part of the floodplain. At the same time, previously stored sediment is released from storage when channel deposits are reworked and bank erosion allows former floodplain deposits to re-enter the channel.

Say there is an increase in sediment output at the downstream end of the reach over a certain period of time. This might indicate that there has been an increase in rates of primary erosion, resulting in greater inputs of sediment to the channel. Alternatively, there could have been a net release of sediment from storage. This might be caused by channel incision along the reach, or by increased rates of channel migration. Unless these

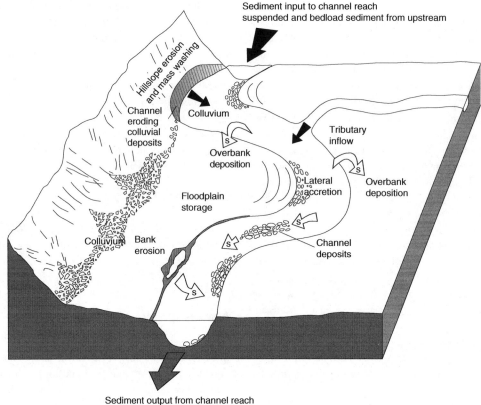

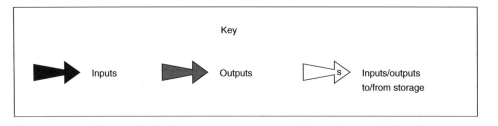

Figure 5.7 The components of a sediment budget for a reach of channel.

changes were observed in the field, or the various inputs, outputs and changes in storage were actually quantified, it would be hard to tell what was happening.

Constructing a sediment budget

By monitoring these different components, a sediment budget can be constructed. For example, rates of

primary erosion can be assessed using the methods described in Chapter 4, pp. 49–51. Inputs from upstream can be quantified by monitoring bedload and suspended sediment transport. It is possible to assess changes in storage from measurements of rates of bank erosion, changes in the number and extent of channel bars, and rates of sediment build-up on the floodplain. Estimates can then be made of the volumes of sediment involved. The sediment budget provides the organising framework for calculating inputs, outputs and stores over a given period of time. Separate budgets can be constructed for coarse and fine sediment.

A relatively recent technique is **sediment finger-printing**, which allows the original source of sediment to be traced, by examining its physical and chemical characteristics. These include mineralogic, mineral-magnetic, chemical, organic, radiometric, isotopic and physical properties (Collins *et al.*, 1997). If several of these are used in combination, unique 'fingerprints' can be defined for sediment originating from different source zones. It is therefore possible to determine the relative contributions from different source zones by comparing these with the 'fingerprints' of sediment particles arriving at the basin outlet.

Applications

The sediment budget approach allows qualitative observations of processes to be integrated with quantitative data on rates of operation. Sediment budgets can take many different forms, although one of the most common is the use of flow diagrams to describe the relationships between sources, stores and transport processes. A classic example is the work of Trimble (1983), who examined changes in sediment budgets in response to land use change in a drainage basin in Wisconsin, USA. This is described in Box 5.2, and demonstrates the very significant influence that sediment storage and remobilisation have on sediment yield.

As well as examining past impacts of human activity, sediment budgets can be used to inform land

---

Box 5.2

USING SEDIMENT BUDGETS TO ASSESS THE IMPACTS OF LAND USE CHANGE IN THE COON CREEK DRAINAGE BASIN

Coon Creek is a 360 km^2 drainage basin in Wisconsin, United States, with a history of land use change. To examine the effects of these changes, Trimble (1983) constructed sediment budgets for two time periods. From 1850 to 1938, severe soil erosion occurred as a result of the introduction of European farming practices and poor land management. This was particularly severe in the upland areas. Between 1938 and 1975 soil conservation measures significantly reduced upland soil erosion. However this did not have any significant effect on the sediment yield, which was 75 $m^3 km^{-2} y^{-1}$ for the first time period and 72 $m^3 km^{-2} y^{-1}$ for the second. (Here the yield is expressed as a volume in m^3 rather than a mass in tonnes.)

Using the sediment budget approach, Trimble concluded that most of the material that was eroded during the first period had ended up in storage. Much of the eroded material was stored on the lower slopes as colluvium and on the floors of valleys. Only a very small proportion of the sediment removed by primary erosion – 5.5 per cent – actually reached the outlet.

During the second period, 1938-75, this stored material was remobilised. This counteracted the reduction in sediment produced by hillslope erosion and meant that there was no change in yield.

Figure 1 shows sediment budgets for the two time periods. On the left hand side of each flow diagram the sources of sediment are shown. The percentages show the relative contributions from each source, the total being 100 per cent. On the right hand side of each diagram are the sediment sinks, or stores. The percentages refer to the proportion of the total eroded sediment that is deposited in each store. The remaining sediment exits the drainage basin.

This research illustrated the importance of sediment storage and remobilisation as a control on sediment yield. Using a sediment budget approach to quantify the inputs, transfers, storage and outputs of sediment, it is possible to gain considerable insight into the operation of sediment systems.

(Continued)

Box 5.2

USING SEDIMENT BUDGETS TO ASSESS THE IMPACTS OF LAND USE CHANGE IN THE COON CREEK DRAINAGE BASIN — CONT'D

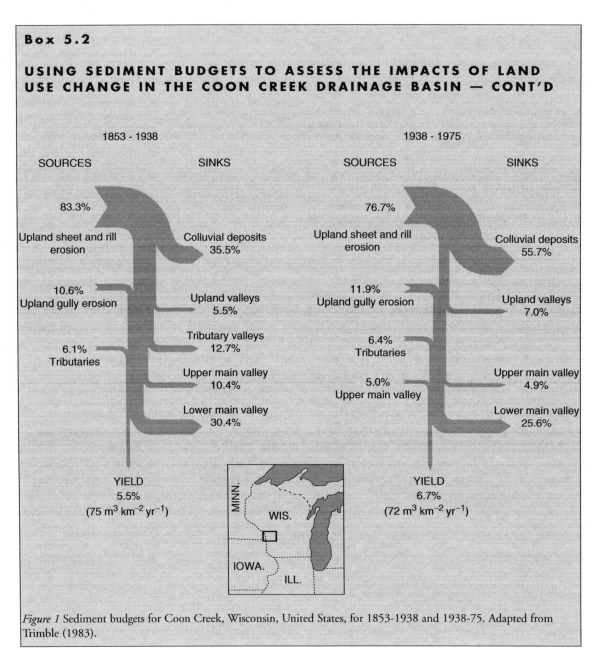

Figure 1 Sediment budgets for Coon Creek, Wisconsin, United States, for 1853-1938 and 1938-75. Adapted from Trimble (1983).

management decisions, assess future impacts of climate change and examine the influences leading to morphological change.

HUMAN ACTIVITY AND SEDIMENT YIELD

Human activity has modified the transfer, storage and yield of sediment in many drainage basins worldwide. There are numerous examples of the dramatic impacts that activities such as deforestation, land use change, mining and dam construction have had on sediment fluxes. These changes affect channel form and behaviour and can have serious implications for channel management. The impacts of such activities are often propagated over a much wider area and the recovery of fluvial systems may take hundreds of years.

Land use change

Various types of land use change can lead to accelerated soil erosion (see pp. 47–49). Sediment supply is also increased by the development of urban areas. During the construction phase, large volumes of material are excavated for roads and buildings.

A number of off-site impacts result from the subsequent transfer of this sediment through river systems. For example, fine sediment often builds up in river channels, making regular dredging necessary. Ecosystems are adversely affected when fine sediment infills the spaces within river gravels at the channel bed. These voids provide a habitat for many species of invertebrates and are important spawning grounds for fish such as salmon and trout. The resultant loss of habitat can seriously reduce ecological diversity. Suspended sediment from agricultural land is the single largest source of pollution in the United States (Lee *et al.*, 1985). It leads to discolouration and can damage the infrastructure of water supply networks. Water treatment to remove suspended sediment is an expensive and time consuming process.

The longer-term effects of land use change may be seen from fluvial records preserved in valley deposits. For example, Starkel (1991a) has identified distinct phases of accelerated valley floor sedimentation in Central European valleys. These correspond to periods of deforestation and land use change during the late Bronze Age, and more extensively during Roman times. Widespread alluviation of valley floors also occurred during the medieval period as a result of agricultural development (Knox, 1995). The effects of land clearance are easier to identify in the New World, where European farming techniques were introduced relatively recently. A layer of 'post-settlement alluvium' is commonly found in agricultural landscapes. These valley floor deposits are mostly composed of fine-grained sediment (Macklin and Lewin, 1997).

Mining

Mining involves the extraction of fossil fuels, metal ores (lead, zinc, copper), rare metals and aggregates (sand and gravel). Mining operations generate huge volumes of waste sediment which, if poorly managed, can easily enter rivers. This includes everything from large fragments of rock to fine sediments and slurries. Before the introduction of legislation in Europe and North America, mine waste was discharged directly into rivers. In many parts of the world this practice continues, affecting whole river systems.

Coarse sediment builds up in river channels and increases the channel bed elevation as well as reducing channel capacity. This increases the risk of flooding during high flows, especially when the raised channel bed is higher than the surrounding floodplain. Channels also become unstable, with frequent shifts in position. Despite the increase in sediment supply, there is a limit to the amount of coarse sediment that can be transported by the receiving rivers. Much of this material is deposited along the way, building up vast stores of sediment on floodplains and valley floors. Long after mining operations have ceased, this sediment continues to be reworked, meaning that recovery is a long and complex process (Box 5.3).

Further problems are associated with the widespread dispersal of highly toxic substances. Heavy metals, such as zinc, copper and cadmium are easily adsorbed to charged sites on the surface of silt and clay particles. Pollutants are then carried with the suspended load and dispersed over a wide area. This is a long-term problem, because deposition

Box 5.3

CASE STUDY: IMPACTS OF GOLD MINING IN THE SACRAMENTO VALLEY, CALIFORNIA

Gold mining has had long-lasting effects in the basin of the Sacramento River, California. Large-scale gold mining resulted in the removal of over 1 billion m³ of sediment between 1853 and 1884 (Gilbert, 1917). These hydraulic mining operations, carried out along the tributaries of the main river, involved directing high pressure hoses on to the valley walls to remove sediment. This was then processed through large sluices to separate gold deposits from gravel.

During the 1850s much of this sediment remained in the tributaries. However, in 1862 large floods delivered huge quantities of mining-derived sediment to the main channels in the lower Sacramento Valley with serious consequences for the growing agricultural communities there. In subsequent years, farms and towns would frequently become inundated with sediment during local flood events. In response, numerous private levees were built in an attempt to contain floodwaters within the channels. However, these exacerbated the problem, concentrating flows and making downstream flooding worse. In addition, levees were frequently breached by sediment-laden flood flows. So much sediment was carried by the channels that they aggraded rapidly, raising bed elevations and increasing the incidence of flooding. As well as the impacts on farms and towns in the valleys, the large volumes of sediment meant that it became increasingly difficult for river traffic to navigate upstream. Eventually, as a result of increasing political and legal pressure from the farming community, hydraulic mining operations ceased.

The mining sediment problem on the Yuba (a tributary of the Sacramento) and Sacramento Rivers was examined by the eminent geologist and geomorphologist G.K. Gilbert. The findings of this study were published in 1917 and are still widely cited today. Gilbert developed a conceptual model of sediment transport which is known as the **sediment wave model**. This refers to the wave of sediment that is generated in response to an increase and subsequent decrease in sediment supply:

> The downstream movement of the great body of debris is analogous to the downstream movement of a great body of storm water ... The debris wave differs from the water wave in the fact that part of its overflow volume is permanently lodged outside the river channel ...
>
> Gilbert (1917: 31)

On the basis of this model, and measurements of the river bed elevation made over a period of time, Gilbert predicted that the rivers would return to their previous elevations by 1967. Subsequent work has shown that, by the 1960s, channel bed elevations had decreased to pre-mining levels, and it might be assumed that sediment loads have returned to pre-mining levels. However, a more detailed analysis carried out over a number of years by James (1999) has shown that these rivers have not in fact 'recovered' from the effects of hydraulic mining. Detailed surveys of affected streams have shown that a considerable volume of sediment is stored in river valleys. Dramatic aggradation led to the deposition of huge volumes of sediment on floodplains. Subsequent channel incision has left this material behind, sitting high above the present river channels in vast sediment reservoirs. Large quantities of sediment also lie in storage on floodplains and beneath levees. This sediment is still being remobilised today, adding to the sediment load of these tributary streams (James, 1999). A large quantity of sediment has yet to be moved. In terms of human time scales, the sediment regime has been permanently altered.

leads to the contamination of floodplains and other stores of fine sediment. Reworking of this material remobilises contaminated sediment, allowing the problem to persist for long periods of time. Pollution by cadmium, lead and zinc on the floodplain of the River South Tyne in northern England has retarded the re-establishment of bankside vegetation (Macklin and Lewin, 1989).

The mining of aggregates (sand and gravel) from the bed of river channels can lead to serious erosion problems, both upstream and downstream. If sediment is removed at a faster rate than it is replaced, there is less sediment to transport, but no reduction in energy. The excess energy, which has to be expended somehow, is then used to erode the bed and banks. Channels can become deeply incised with steep, unstable banks. Associated management problems include the undermining of bridges and other structures.

Dam construction

Impacts

Dams disrupt the continuity of river systems, trapping all bedload and most of the suspended load. Dams on the highly regulated Ebro River in north-east Spain trap 95 per cent of the suspended load (Vericat and Batella, 2005). Dams can present serious management problems downstream, creating sediment-starved flows of 'hungry water'. For example, the Colorado River has incised (deepened) its bed by up to 6 m downstream from the Davis Dam (Williams and Wolman, 1984). Severe channel degradation has also occurred on the Arno River in Italy, where gravel mining has exacerbated sediment losses caused by dams. Rates of incision are between 2 m and 5 m on average, and up to 9 m in places (Rinaldi and Simon, 1998). Further downstream, aggradation problems are often associated with the accumulation of this eroded sediment. However, the Aswan High Dam, on the River Nile in Egypt, has reduced the sediment yield to such an extent that the Nile Delta, 1,000 km downstream, is receding by 150 m a year due to coastal erosion.

Reservoir sedimentation

The sediment that builds up in reservoirs is a major determinant in the 'design life' of a dam. Sediment reduces reservoir storage capacity and the annual loss to global reservoir storage is between 0.5 per cent and 1 per cent (WCD, 2000). For many reservoirs, though, high sediment yields mean that annual storage depletion rates are as high as 4 per cent or 5 per cent. This means that the majority of the reservoir storage capacity is lost after only twenty-five or thirty years. Worldwide there are numerous reservoirs that have had to come out of commission because they have become completely infilled with sediment. Even when only a small percentage of the storage volume has been lost, operational problems start to occur.

CHAPTER SUMMARY

Sediment that has been generated by primary erosion is transferred from the hillslopes to headwater streams and onwards through the channel network. A certain proportion exits the drainage basin, although a significant volume is deposited along the way, entering permanent or temporary storage. The volume of sediment that leaves the basin outlet in a certain period of time is called the sediment yield. This is commonly expressed in terms of the specific sediment yield, or yield per unit area of the drainage basin. Specific sediment yield is controlled by a number of factors, including drainage basin size, topography, climate, erodibility and human activity, and considerable variations are seen on a global scale. Basin size is an important control because the available storage volume is much greater for large drainage basins. As a result, the specific sediment yield generally decreases with increasing basin area. There are important differences between coarse and fine sediment transfer. Fine material is carried in suspension, while coarse sediment is transported along the bed of the channel. The finest component of the suspended load, the wash load, can be transported over considerable distances in a short time. In contrast, bed load transport is much more localised. Human activity has significantly modified the transfer, storage and yield of sediment in many drainage basins worldwide. Land use changes often increase rates of erosion, while huge volumes of sediment are produced by mining activity. This leads to a build-up of sediment in river channels and increased deposition on floodplains and valley floors. Further impacts are associated with the later release of this sediment from storage.

Rivers are starved of sediment as a result of dam construction, and the mining of sand and gravel from river channels. Incision often results, with channels sometimes deepening their channels by several metres. Storage effects mean that the response to human activity is highly complex, and adjustment to these changes can take decades or centuries.

FURTHER READING

Introductory texts

Mount, J.F., 1995. *California Rivers and Streams.* University of California Press, Berkeley and Los Angeles, CA. Includes several interesting and very readable chapters about the impacts that mining, logging, urbanization and dams have had on California's rivers.

Woodward, J. and Foster, I.D.L., 1997. Erosion and suspended sediment transfer in river catchments. *Geography*, 82(4): 353–76. This clearly written paper includes a number of useful case studies and is highly recommended.

Selected journal articles

Church, M. and Slaymaker, O., 1989. Disequilibrium of Holocene sediment yield in glaciated British Columbia. *Nature*, 337: 452–4. A significant paper highlighting the lasting legacy of the most recent Pleistocene glaciation on sediment yield.

Milliman, J.D. and Syvitski, J.P.M., 1992. Geomorphic/tectonic control of sediment discharge to the ocean: the importance of small mountainous rivers. *Journal of Geology*, 100: 525–44. Explains the significance of drainage area and topography, and includes a table of data for 280 rivers.

Techniques

Kondolf, G.M. and Piégay, H. (eds), 2003. *Tools in Fluvial Geomorphology.* John Wiley & Sons, Chichester. Suspended sediment sampling and modelling are discussed in the chapter by Hicks and Gomez. There is also a detailed discussion of sediment budgets by Reid and Dunne.

Website

World River Sediment Yields Database, www.fao.org/ag/aGL/aglw/sediment/default.asp. Data on annual sediment yields for world rivers and reservoirs. Searchable by country, continent and river. Bear in mind that the information has come from a number of different sources, some of which are more reliable than others.

6

FLOW IN CHANNELS

The water in river channels moves down slope under the influence of gravity, which causes it to deform and flow. This movement is resisted by frictional forces between the flow and channel boundary, and within the flow itself. In this chapter you will learn about:

- The forces driving and resisting flow.
- How the flow in a river varies over space and time.
- Types of flow resistance.
- Flow behaviour.

INTRODUCTION TO FLOW IN RIVER CHANNELS

Forces driving and resisting the flow of water

A force is anything that moves an object, or causes the speed or direction of a moving object to change. Forces are vector quantities, which means that they have both magnitude (size) and direction. The unit of force is the newton (N), and force magnitude is defined by the mass of the object and the acceleration produced.[1] Forces are always mutual. In other words, if a force is exerted on an object, the object will react with an equal and opposite force.

In most situations, several forces are involved, so the balance between driving and resisting forces is usually considered. Forces acting on an object are balanced if the object is stationary, or if it is moving at a constant velocity. The driving force causing water to flow (whether in a channel, rill, gully or overland) is the down-slope component of gravity. This acts on a given mass of water, causing it to deform (flow) and move in a downstream direction over the channel boundary (bed and banks).

Opposing this movement are resisting forces. Resistance occurs because of friction between the flow and channel boundary. Also, the fluid itself resists deformation because of internal forces within the flow.

As water moves down slope, it exerts a shearing force, or shear stress, on the channel boundary (shear stress is represented by the Greek letter tau, τ). The **bed shear stress (τ_0)** is expressed as a force per unit area of the bed (in N m^{-2}) and increases with flow depth and channel steepness. This relationship is described by the du Boys equation (Box 6.1).

Channel parameters

In order to describe the flow of water in river channels it is necessary to define some basic channel parameters, most of which are illustrated in Figure 6.1. Channel size can be defined by its cross-section: a slice taken across the channel, perpendicular to the direction of flow.

Box 6.1

THE DU BOYS EQUATION

The shear stress acting on the bed of a channel is defined by:

$$\tau_0 = \rho ghS$$

where τ_0 is the spatially averaged bed shear stress, ρ is water density, g is the acceleration due to gravity, h is the depth of flow and S is the slope.

The area of the cross-section is given by the product of channel width and the mean flow depth. At a given cross-section, the cross-sectional area changes through time in response to fluctuations in discharge (defined in previous chapters). The maximum discharge that can be contained within the channel, before water starts to inundate the floodplain, is called the **bankfull discharge**. The width of the channel at bankfull discharge is called the **bankfull width**. It should be noted that there are several issues associated with the definition of bankfull discharge for many river systems (see Chapter 3, p. 32).

The shape of a river channel affects its hydraulic efficiency, something that can be quantified by calculating the **hydraulic radius**. This is a measure of how much contact there is between the flow and channel boundary, and is calculated from:

$$\text{Hydraulic radius} = \frac{\text{Cross-sectional area (m}^2)}{\text{Wetted perimeter (m)}}$$

The **wetted perimeter** is the length of channel boundary that is in direct contact with the flow at a given cross-section. An example is provided in Figure 6.2, which shows two channel cross-sections. For the purposes of this illustration, it will be assumed that the only difference between them is their shape, channel A is wide and shallow, while channel B is narrow and deep. Both have the same cross-sectional area but the wetted perimeter is larger for channel A, resulting in a lower hydraulic radius. Assuming all else is equal, the loss of energy arising from friction with the bed and banks will be greater for channel A. Channel B is therefore more hydraulically efficient. For wider channels, the hydraulic radius is very similar to the flow depth.

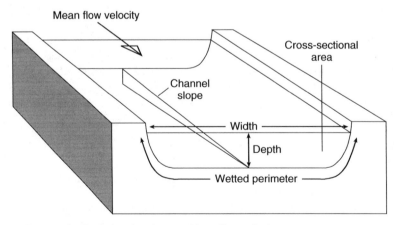

Discharge = Cross-sectional area x Mean flow velocity
Hydraulic radius = Cross-sectional area/wetted perimeter

Figure 6.1 Basic channel parameters. After Summerfield (1991).

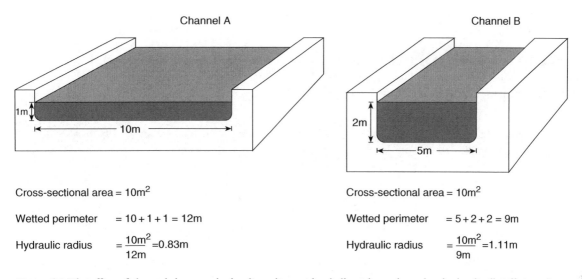

Channel A

Cross-sectional area = 10m^2

Wetted perimeter   = 10 + 1 + 1 = 12m

Hydraulic radius $= \dfrac{10m^2}{12m} = 0.83m$

Channel B

Cross-sectional area = 10m^2

Wetted perimeter   = 5 + 2 + 2 = 9m

Hydraulic radius $= \dfrac{10m^2}{9m} = 1.11m$

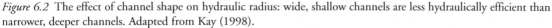

Figure 6.2 The effect of channel shape on hydraulic radius: wide, shallow channels are less hydraulically efficient than narrower, deeper channels. Adapted from Kay (1998).

Channel slope is usually expressed as a gradient (difference in channel bed elevation along a given length of channel in meters divided by that length in metres). This is related to, but not necessarily the same as, the **water surface slope**, the downstream change in water surface elevation along the channel. Water surface slope is an important variable because it closely approximates the **energy slope** along a particular length of channel. As water flows through the channel, potential energy is converted to kinetic energy. This is in turn converted to heat energy, which is generated as a result of friction,[2] and 'lost' from the channel. As a result there is a downstream reduction in the total energy 'possessed' by a given parcel of water. The steepness of the energy slope reflects the rate at which energy is being expended. Further detail on energy relationships is provided in Box 6.2.

Flow velocity

Flow velocity varies over both space and time in natural channels. It is determined mainly by the channel slope, roughness and cross-sectional form (remember that channel depth and cross-sectional area change with discharge). If you have ever waded out into a stream, you will know that the flow velocity, like the depth of flow, tends to increase as you move out into the channel. This is because of friction between the flow and the channel boundary, which is greatest near the bed and banks. Together with the effects of turbulence, the effects of this frictional resistance create variations in velocity distributions that are seen at different spatial and temporal scales. These are briefly discussed below.

- *Variations with time.* At any given point within the flow, the velocity fluctuates rapidly because of the effects of turbulence. This means that instantaneous velocities at a specific location can be much higher or lower than the **time-averaged velocity** that is recorded by a flow meter (described in Box 3.1). Over periods of days, weeks or months, variations in velocity are also seen at the channel scale in response to discharge fluctuations.
- *Variations with depth.* These can be seen from measurements of time-averaged velocity made at different vertical heights above the channel bed (imagine a vertical line stretching upwards from a specific point on the channel bed). An example of a **vertical velocity profile** is shown in Figure 6.3a.

Box 6.2

ENERGY RELATIONSHIPS AND THE BERNOUILLI EQUATION

Potential, kinetic and pressure energy

Energy is the capacity to do **work**. Work in this sense has a specific meaning and is used to express the forces required to set an object in motion. In river channels, work is done in overcoming flow resistance and in moving grains of sediment. The energy required to carry out this work is stored in three main ways: as potential energy, as kinetic energy and as pressure energy.

- **Potential energy** is stored energy. Water falling as precipitation over the headwaters of a drainage basin has potential energy by virtue of the fact that it has been lifted to that elevation by atmospheric processes. Potential energy is usually expressed as an **elevation head** (in metres) above a reference datum, such as sea level, and increases with elevation above that datum.
- **Kinetic energy** is possessed by moving objects and was mentioned in Chapter 3 in the context of soil erosion by raindrop impact. When a 'parcel' of water flows downhill, its potential energy decreases as it is converted to kinetic energy. The kinetic or **velocity head** can be derived by dividing kinetic energy by the weight of water to give a head in metres.*
- **Pressure energy** is related to the depth of water at a given point in the channel. It is expressed as the **pressure head** at the channel bed – in other words the depth of water (in metres).

At any point in the channel, the total energy 'possessed' by the water is the sum of potential energy, kinetic energy and pressure energy (Figure 1). This can be written as an equation – the **Bernoulli Equation**. This was derived by the eighteenth century Swiss mathematician Daniel Bernoulli and is widely used in hydraulics. For flow in open channels the equation states that:

Total energy = Potential energy + Kinetic energy + Pressure energy

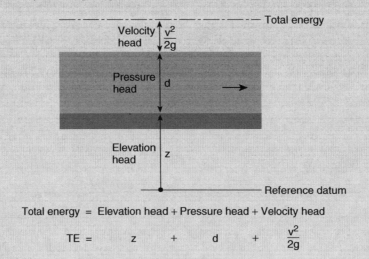

Total energy = Elevation head + Pressure head + Velocity head

$$TE = z + d + \frac{v^2}{2g}$$

Figure 1 The total energy at a given point in a channel is the sum of the elevation head, pressure head and velocity head at that point. Adapted from Kay (1998).

Box 6.2

ENERGY RELATIONSHIPS AND THE BERNOUILLI EQUATION—CONT'D

Or, to express the total energy head in metres:

$$Total\ head = Pressure\ head + Velocity\ head$$
$$+ Elevation\ head$$

$$TE = d + \frac{v^2}{2g} + z$$

where TE = total energy head (m), d = water depth (m), v = flow velocity (ms⁻¹), g = gravitational constant (9.81 ms⁻²) and z = elevation head (m).

The rate at which potential energy is converted to kinetic energy along a given length of channel depends upon the loss in elevation, assuming there is no change in flow depth. The rate of conversion from potential to kinetic energy is greater for a steeply sloping reach than for a reach of the same length but with a gentler gradient. Most of the kinetic energy is in turn converted to heat energy as a result of boundary and internal friction – you can observe this frictional heating effect by rubbing your hands together. The heat energy does not result in an appreciable warming of the water, as it is rapidly dissipated throughout the

flow and 'lost' to the atmosphere. A very small proportion of the total energy is converted to sound energy – the various noises made by flowing water – although this is miniscule by comparison.

Along a length of channel, these energy 'losses' result in a downstream reduction in the total head. This is illustrated in Figure 2, which shows the total head at the upstream (A) and downstream (B) ends of a channel and the **head loss** that is observed between them. The **energy continuity equation** therefore needs to account for this expenditure of energy along the channel:

Total energy at A = Total energy at B + Head loss

Of particular significance for flow and sediment movement is the way in which the velocity and pressure heads change in relation to each other. Over short lengths of channel (a few metres), the elevation head can be considered to be the same and frictional head losses can be ignored. Say there is an increase in the depth of flow along the channel. In order to ensure flow continuity, there must be a corresponding

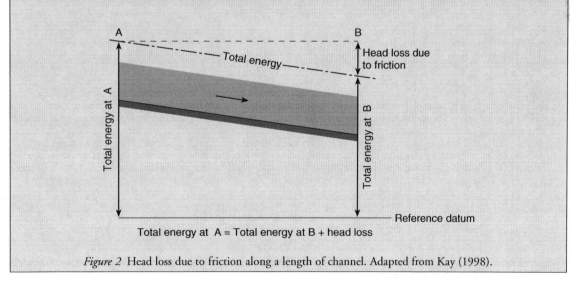

Figure 2 Head loss due to friction along a length of channel. Adapted from Kay (1998).

Box 6.2

ENERGY RELATIONSHIPS AND THE BERNOUILLI EQUATION—CONT'D

decrease in the mean flow velocity. Similarly, a reduction in depth would result in an increase in flow velocity. This is rather different from the way in which crowds of people move (Figure 3). When people move through a constriction, they tend to slow down as they move closer together. This causes a tail-back behind the constriction. Water does not back up in the same way. Instead, it flows more quickly through the constriction in order to ensure flow continuity. Just upstream from the constriction there may be an increase in flow depth. This creates an increased water surface slope and is associated with the increased energy gradient needed to move water through the constriction. Since the velocity head increases, there must be a corresponding decrease in the pressure head. This is called the **Bernoulli principle**. It is the same principle that provides some of the lift that allows an aeroplane to fly through a different fluid: air. The shape of the aircraft wing means that air moving over the top of it has to travel further, and faster, than the air flowing under it. The increase in velocity over the upper surface of the wing results in a reduction in pressure above it. This contributes some of the lift that enables the plane to stay airborne. The same principle also causes a lift force to be exerted on sediment grains on the channel bed (Chapter 7).

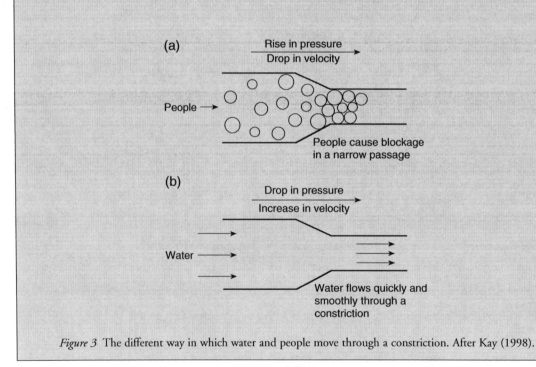

Figure 3 The different way in which water and people move through a constriction. After Kay (1998).

*Kinetic energy $= \frac{1}{2}mv^2$ where: m = mass and v = velocity. In order to express this as a head in metres, the kinetic energy is divided by weight (mass $\times$ g), which removes the mass term from the equation to give the kinetic head (in m): *kinetic head* $= \frac{v^2}{2g}$.

(a)

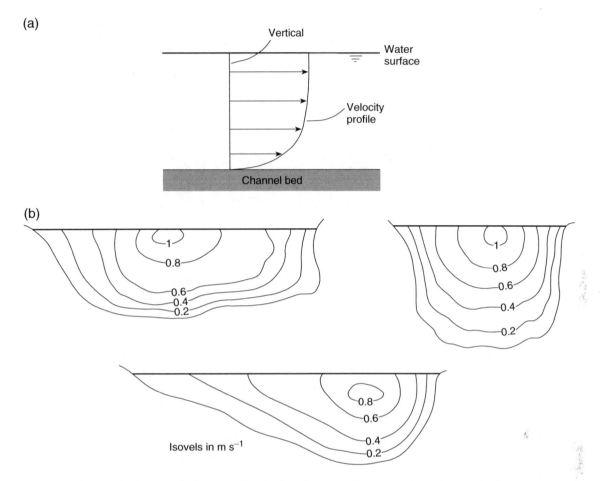

Figure 6.3 (a) Example of a vertical velocity profile, showing changes in time-averaged velocity with depth.
(b) Channel cross-sections illustrating the effect of channel shape on the velocity distribution. (b) is after Knighton (1998).

At the bed itself, the velocity is zero, but increases with vertical distance above the bed. The actual *rate* of increase, or **velocity gradient**, is greatest close to the bed, levelling off further away from the bed. The vertical velocity gradient at any point determines the shear stress exerted on the bed at that point.

- *Variations across the cross-section.* Velocity profiles for three different cross-sections are shown in Figure 6.3b, where **isovels** – lines of equal velocity – have been plotted. As can be seen, the fastest flow

occurs towards the centre of the channel. At this cross-sectional scale, the average flow velocity can be calculated by making a number of measurements of velocity across the channel and at different depths. A description of how to do this is provided by Goudie (1981).

- *Downstream variations.* Although there is typically a decrease in channel slope along the length of a channel, the velocity generally shows little change or increases slightly. This is because the decrease in slope is often compensated for by a downstream

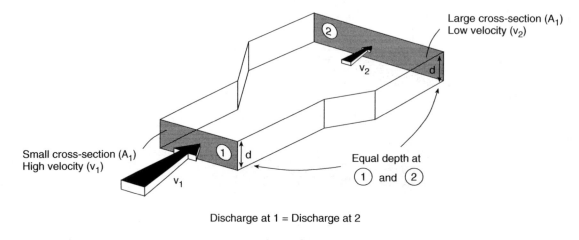

Figure 6.4 Illustration of the concept of flow continuity.

decrease in channel roughness and an increase in hydraulic efficiency. The effects of slope, roughness and hydraulic efficiency are discussed later in this chapter. Downstream changes in velocity will be considered in Chapter 8.

The concept of flow continuity

A casual observer walking alongside a natural stream channel might notice that the deeper sections are relatively slow flowing, while the shallow sections are relatively fast-flowing. The reason for this is that – assuming no tributaries join the channel and there are no significant interactions with groundwater – the same volume of water has to travel through each section in a given time. If this did not happen, the flow would start building up in some parts of the channel. Other parts would run dry as water flowed downstream faster than it was supplied from upstream.

The concept of flow continuity is illustrated in Figure 6.4, which shows a length of artificial, watertight channel with a constant discharge entering at the upstream end. The discharge passing through cross-section (1) at

the upstream end of the reach is equal to the discharge flowing through cross-section (2) at the downstream end. Since discharge is the product of cross-sectional area and flow velocity, it follows that:

$$\text{Discharge at (1)} = \text{Discharge at (2)}$$

$$A_1 \times v_1 = A_2 \times v_2$$

where A_1 and A_2 = cross-sectional area at (1) and cross-sectional area at (2); v_1 and v_2 = mean flow velocity at (1) and (2). This is known as the **volumetric continuity equation**. From the example given in Figure 6.4 you can see that although the upstream cross-sectional area at (1) is smaller than the downstream one at (2), the upstream velocity is faster, so the same volume of water passes through the smaller cross-section over a given period of time.[3]

Following on from this is the concept of **conservation of mass**. The mass of a given volume of water can be calculated by multiplying that volume by its density. According to the volumetric continuity equation above, the volume of flow does not change and, since water cannot be compressed, its density (1 kg per litre) does

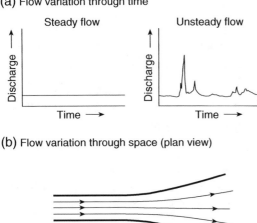

(a) Flow variation through time

Steady flow   Unsteady flow

Figure 6.5 (a) Flow variation through time: steady and unsteady flow. (b) Flow variation through space: uniform flow and non-uniform flow.

not change either. Therefore the mass of water passing (1) is equal to the mass of water passing (2). If this was not the case, water would be spontaneously created or destroyed somewhere along the channel.

Variations through time: steady and unsteady flow

Steady and unsteady flows are classifications of flow variations through time. In the example above, it was assumed that the discharge entering the reach did not change through time, something that is called **steady flow**. In natural channels, the flow is usually **unsteady**, varying through time as the drainage basin responds to inputs of precipitation (Figure 6.5a).

Variations through space: uniform and non-uniform flow

Flows can also be classified according to variations over space. In a channel reach with a constant slope and cross-sectional shape, there will be no variation in either depth or velocity along the reach. This is called **uniform flow** and occurs in the upstream segment of the channel illustrated in Figure 6.5(b). The uniformity of the flow is indicated by streamlines – lines indicating the mean direction of individual 'parcels' of flow – which are parallel. Although most hydraulic equations are based on the assumption that the flow is uniform, this is rarely the case for natural channels, where the shape and dimensions of the channel vary in a downstream direction. There are also bends and obstacles to flow such as constrictions and channel bars. The flow expands into wider sections and becomes concentrated where there are constrictions. This means that the streamlines are no longer parallel, and the flow is described as **non-uniform** (Figure 6.5b). It is only under uniform flow conditions that the channel bed slope, water surface slope and energy slope are the same.

There are two types of non-uniform flow. **Gradually varied** flow is illustrated in Figure 6.5(b) and reflects changes that occur over distances of tens of metres or more. **Rapidly varied** flow is associated with sudden changes in channel width, depth or alignment. In these situations, the streamlines cannot follow the line of the channel and something called flow separation occurs. Hydraulic jumps and drops are also associated with rapidly varied flow. These types of flow behaviour will be discussed in the section after next.

FLOW RESISTANCE

A surprising amount of energy has to be used by flowing water in order to overcome flow resistance. It has been estimated that as much as 95–97 per cent of the total energy of a river is expended in this way (Morisawa, 1968). **Flow resistance formulae** express the relationship between flow velocity, channel slope, roughness and cross-sectional shape. Velocity increases with channel slope, but decreases with increasing boundary roughness. For example, a concrete-lined channel offers much less frictional resistance than a rocky, boulder-strewn channel. The hydraulic radius is also significant, since this determines the area of contact between the flow and boundary. Roughness is difficult to measure directly, so resistance formulae include an empirically-derived **friction coefficient** (examples are given in Box 6.3).

Box 6.3

FLOW RESISTANCE FORMULAE

The Chézy equation is named after the eighteenth century French hydraulic engineer, Antoine de Chézy. This was later refined by the nineteenth century Irish engineer, Robert Manning. The Darcy–Weisbach equation has a long history of development and is named after two of the great hydraulic engineers of the nineteenth century. It has a sounder theoretical basis than the Manning and Chézy equations, although the Manning equation is still widely used today.

Chézy equation

$$v = C\sqrt{Rs}$$

Manning equation

$$v = \frac{R^{0.67} s^{0.5}}{n}$$

Darcy–Weisbach equation

$$v = \sqrt{\frac{8gRs}{f}}$$

where v = velocity, C = Chézy roughness factor, R = hydraulic radius, s = channel slope, n = Manning roughness coefficient, g = acceleration due to gravity (9.8 m s^{-2}) and f = Darcy–Weisbach friction factor.

The Chézy coefficient (C) represents gravitational and frictional forces. Its value decreases with increasing roughness. Manning's roughness coefficient (n) is usually determined from tables. (Table 1 gives some values of Manning's 'n' for natural channels.) Another method is to use photographs to make comparisons with channels of known roughness.

Example application of the Manning equation:

Calculate the velocity of a lowland meandering channel with riffles and pools, which has a slope of 0.001 m m^{-1}, a wetted perimeter of 9 m and a cross-sectional area of 10 m^2.

R = 1.11 m (cross-sectional area/wetted perimeter) and 'n' = 0.040 (from Table 1), so:

$$v = \frac{1.11^{0.67} \times 0.001^{0.5}}{0.040} = 0.85 \text{ m s}^{-1}$$

Problems and limitations

Although widely used, these formulae have limitations. One of the main problems is that roughness is controlled by a number of different factors, including bed material size, bedforms and vegetation (pp. 78–80). This cannot be adequately represented by a single, empirically derived roughness coefficient. Flow resistance also changes with stage, being highest at low flows and lowest at bankfull stage. Once overbank flow starts to occur, the increased roughness of the floodplain surface greatly increases the overall flow resistance (see pp. 88–91).

Table 1 Some values of Manning's 'n' for natural channels

Channel description	Minimum	Normal	Maximum
Small channels (width < 30 m)			
Lowland channels:			
Unvegetated, straight channels	0.025	0.030	0.033
Unvegetated winding channels with some pools and shallows	0.033	0.040	0.045
Winding channels with vegetation and stones on bed	0.035	0.045	0.050
Sluggish vegetated channels with deep pools	0.050	0.070	0.080
Heavily vegetated channels with deep pools	0.075	0.100	0.150
Mountain streams (with steep banks and no in-channel vegetation):			
Mainly gravels and cobbles with few boulders	0.030	0.040	0.050
Cobbles with large boulders	0.040	0.050	0.070
Large channels (width > 30 m)			
Regular channels with no boulders or vegetation	0.025	–	0.060
Irregular channels	0.035	–	0.100

Source: Adapted from Chow (1959).

Channel resistance

At the valley scale, flow resistance increases when the channel comes into contact with the valley margins. This occurs in confined valley settings and where there are changes in valley alignment. The three-dimensional shape of the channel is also influential, since resistance is increased by irregularities in the banks, downstream changes in cross-section, and where the flow moves around bends. Bedrock-influenced channels can be highly irregular in form, with large variations in slope, width and channel cross-section. The high resistance of such channels is further increased by features such as cascades, vertical steps and potholes which increase form resistance (see below).

In a detailed investigation of variations in total flow resistance for different channel types along the Sabie River in South Africa, Heritage *et al.* (2004) reported extreme values of total flow resistance for bedrock-influenced channel reaches during low flows. These values were calculated for mixed anabranching channel sections, where the flow is divided into a number of separate bedrock-dominated distributary channels under low flow conditions (the distributaries are separated by bedrock core bars that are overlain by cohesive sediment and vegetation). At low discharges, the flow in each distributary is very shallow and the highly fissured bedrock pavement means that the wetted perimeter is very large and tortuous (a large wetted perimeter means a smaller hydraulic conductivity and greater flow resistance). Numerous pools and rapids form within the fissures, with steep water surface slopes and very high rates of energy dissipation. Added to this are the effects of numerous boulders, which create obstacles to the flow. As discharge increases, a decrease in resistance is seen as these features become increasingly submerged by the flow. During flood flows, the vegetated bars separating the distributary channels become inundated, with an increase in resistance that is attributed to the increased resistance of the vegetation (Heritage *et al.*, 2004).

Boundary resistance

There are two components of boundary resistance. The first of these, **grain roughness**, relates to the effects of the individual grains making up the channel boundary. **Form roughness** refers to features such as ripples and dunes, which are created when certain alluvial substrates are moulded by the flow.

Grain roughness

In general terms, flow resistance increases with the diameter of individual grains. However, an important factor is the depth of flow relative to the size of the particles. This can be expressed in terms of a ratio:

$$\frac{d}{D}$$

where d is the flow depth and D is a characteristic grain size index; the median size of the bed sediment is often used. This ratio is used in many process-based equations in fluvial geomorphology and acts as a very significant control on the overall resistance in a channel (Robert, 2003).

Bathurst (1993) compares the ratio of flow depth to characteristic grain size for different channel types along an idealised channel system. For a sand-bed channel, the flow depth may be over a thousand times greater than the diameter of the individual sand grains (2 mm or less). For gravel-bed channels, the ratio may be between 5 and 100, depending on the dominant grain size, which can range from cobbles (up to 250 mm) down to fine gravels (10 mm). In boulder bed channels, where most particles have diameters of 250 mm or more, the particles may project through the whole depth of flow, with a d/D ratio of less than 1. Where the stream bed consists of gravel or cobbles, grain roughness can be the dominant component of flow resistance (Knighton, 1998). However, the effect of grain roughness is often 'drowned out' as the depth increases.

Grain size, and the spacing of individual grains, can also have a significant influence on the structure of turbulent flows. These effects will be discussed in the next section (see section on 'hydraulically rough and smooth surfaces').

Form roughness

In sand-bed channels, it is possible for a wide range of flows to shape the channel bed. At different flow intensities, a sequence of **bedforms** develops (the formation of which will be discussed in Chapter 7). These include dunes, which are scaled to the depth of flow in the channel. Bedforms increase turbulence and can cause flow separation, leading to significant energy losses at high flows. Varying levels of resistance are associated with different types of bedforms (Simons and Richardson, 1966) and in sand bed streams the presence of these forms often exceeds grain roughness in importance (Knighton, 1998). In gravel-bed rivers, longitudinal variations in channel slope and bed roughness are often associated with periodic features called **riffles and pools**. Increased flow resistance occurs mainly as a result of ponding upstream from the shallower riffles (Hey, 1988). These are interspersed by deeper, slower moving pools, with a spacing of between five and seven times the channel width. Channel bars also increase flow resistance, particularly in braided channels. Even at higher flow stages, bars can account for between 50 and 60 per cent of total flow resistance (Prestegaard, 1983). Micro-scale variations in the bed topography of many gravel-bed channels are associated with smaller features called cluster bedforms. These consist of a single protruding obstacle, such as a large pebble, with associated accumulations of finer material immediately upstream and downstream (see Chapter 7). Cluster bedforms have a significant effect on shear stress distributions and flow resistance (Lawless and Robert, 2001).

In steep, rocky channels, sequences of **steps and pools** may form. These are associated with very high rates of energy expenditure, particularly at low flows when considerable energy has to be dissipated in hydraulic jumps and pools (Bathurst, 1993). The extreme resistance reported by Heritage *et al.* (2004) for the Sabie River was associated with bedrock-influenced rapids and cataracts. This morphology is highly irregular in form, with very high energy dissipation caused by hydraulic jumps, constrictions and other disturbances to the flow.

Other controls on flow resistance

Riparian and in-channel vegetation increases flow resistance. This varies with stage, because vegetation that is upright at low flows may become flattened at higher flows. Seasonal effects are also seen when vegetation dies back during the winter months. Patchy growth can lead to considerable variations in resistance across the channel bed. In some channels woody debris builds up to create additional resistance.

Sediment transport may also be of some significance. A high suspended load increases fluid viscosity, reducing turbulence and, in turn, flow resistance (Knighton, 1998). Several studies have investigated the dependence of flow resistance on bedload transport for the coarse bed materials typical of mountain rivers. However these effects appear to be small in relation to other controls (Bathurst, 1993).

FLOW BEHAVIOUR

Subcritical, critical and supercritical flow

The unsteady, gradually varied flow in most natural channels is **subcritical**. However, another type of flow behaviour, **supercritical flow**, is also observed. Within supercritical flows, turbulent mixing is less intense, with less deviation from the main downstream direction of flow. As a result, supercritical flows move rapidly and efficiently through the channel. They may overshoot tight bends and can also be highly erosive (Kay, 1998).

The different types of flow behaviour can be predicted by calculating the ratio between the inertial and gravitational forces. The inertial force (see p. 82) is given by v^2/d; where v is the flow velocity, and d is its depth. The gravitational force is the acceleration due to gravity, g. The ratio between these forces is usually expressed in the form:

$$\mathrm{Fr} = \frac{v}{\sqrt{gd}}$$

where Fr = Froude number, v = velocity, g is gravitational constant and d is depth.

At Froude numbers less than 1 the gravitational forces dominate and the flow is subcritical. Conversely, when the inertial forces dominate, at Froude numbers greater than 1, the flow is supercritical. In rare cases, where the Froude number is equal to 1, the flow is described as being **critical**, or transitional.

Figure 6.6 shows what happens when there are transitions between these two types of flow. A **hydraulic drop**

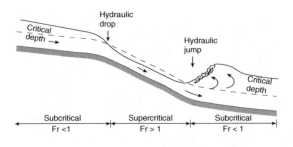

Figure 6.6 Illustration of hydraulic drop and hydraulic jump.

occurs when subcritical flow changes to supercritical flow. In this example, the increase in channel slope increases the flow velocity, resulting in a reduction in depth (the hydraulic drop). This can be seen in Plate 6.1, as water flows smoothly and rapidly down the steep slope of a weir. At the base of the weir, the flow changes back to subcritical, forming a **hydraulic jump**. A breaking wave indicates where this transition occurs. The sudden change in flow conditions at the hydraulic jump is caused by the decrease in slope at the base of the weir. Associated with this is a decrease in velocity and an increase in depth. The high velocity flow has considerable inertia and continues along the bed of the river before it is 'pulled' up to the surface and into the breaking wave. Turbulence is increased in this zone because of shear between the downstream and upstream movements of water.

Flow separation

Flow separation occurs where there are irregularities in the boundary. Examples include abrupt changes in bank orientation, sharp bends and obstructions at the bed such as large boulders. This results in the detachment of the boundary layer, which continues on in the direction of the flow but as a free shear layer. Figure 6.7(a) illustrates flow separation caused by a boulder on the channel bed. In the flow separation zone, between the free layer and the boundary, is a 'bubble' of slow moving, recirculating fluid. The large difference in velocity between the fast moving shear layer and slowly recirculating flow in the separation zone means that large shear stresses develop. The resultant transfer of momentum

leads to the free layer becoming unstable a certain distance downstream from the separation point, where it reattaches to the boundary.

Increased turbulence is created by flow separation and results in a **wake** downstream from the object. Figure 6.7 (b and c) shows the effect of flow separation around an obstruction and at a bend. Since flow separation affects shear stress distributions, it also influences processes of sediment erosion and deposition.

Laminar and turbulent flow

When considering the internal structure of fluid flow, a distinction is made between two quite different types of flow: laminar and turbulent. The British engineer Osborne Reynolds first demonstrated the existence of these two types of flow in his well known experiments on flows through pipes, carried out in the 1870s and 1880s. By injecting a thin stream of coloured dye into the water, Reynolds was able to observe patterns of movement within the flow. At low flow velocities, the dye was seen to travel as a single thread in a straight line through the tube, and was described by Reynolds as *direct* flow (now known as **laminar** or **viscous flow**). In laminar flows, the fluid moves as a series of layers, which slide over one another. This can be visualised as being somewhat similar to the way in which a pack of cards slide over each other when a shear stress is applied. Highly viscous fluids, such as oil or treacle, tend to exhibit laminar flow because of their high resistance to deformation. This can be seen from the 'smooth' way in which these fluids flow over a surface when gently poured. Water has a relatively low viscosity, so laminar flow only occurs at very low flow velocities.

Reynolds found that a second, very different, type of flow occurred at higher velocities. In contrast to laminar flows, a series of horizontal and vertical swirling motions developed, dispersing the dye throughout the flow. Described as *sinuous* by Reynolds, this flow behaviour was subsequently termed **turbulent flow** by Lord Kelvin. Within the three-dimensional body of flow, movement can be in any direction: vertically up or down, sideways, upstream, downstream, or any combination of these. Reynolds found that as flows changed from laminar to turbulent, a transitional flow-type

Plate 6.1 Hydraulic drop and hydraulic jump at a weir on the River Liffey, Country Kildare, Irish Republic. Photograph by Jeanne Meldon.

developed, with the turbulence intensity increasing as the flow became fully turbulent.

From these experiments it was clear that two different types of flow behaviour existed. What was not so clear was how the transition between these flows could be predicted, as velocity is only one of a number of variables that control flow behaviour. Reynolds conducted further experiments using different fluids, and pipes with varying diameters. From these experiments, he derived an equation to define the transition from laminar to turbulent flow as a function of a single parameter, the **Reynolds number (Re)**. At low Reynolds numbers laminar flow occurs and at high values, turbulent flow. The concept of a Reynolds number is fundamental to much

of modern fluid dynamics. It is calculated using the **Reynolds equation**, which expresses the ratio between **inertial** and **viscous forces** acting on the fluid. The inertia of an object – in this case a body of flowing water – is defined by its mass. Inertia determines how difficult it is to set something in motion (here: to initiate flow) but also how difficult it is to stop it, slow it down or change its direction once it has started moving. The greater the mass of water, the more inertia it has. Fluids that are denser than water (e.g. mercury) have more inertia because their mass per unit volume is greater. The inertial forces also increase with velocity.

Acting against the inertial forces are viscous forces, which resist fluid deformation and flow. The viscosity of

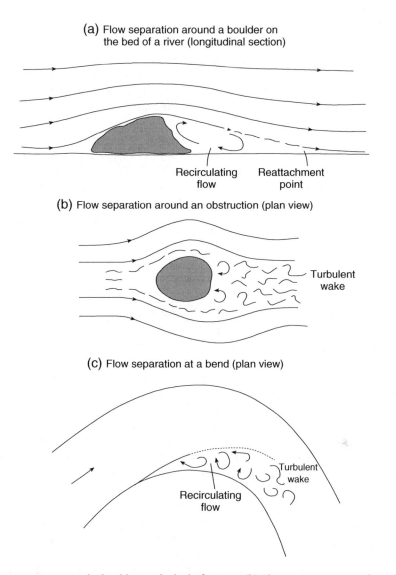

(a) Flow separation around a boulder on the bed of a river (longitudinal section)

Recirculating flow

Reattachment point

(b) Flow separation around an obstruction (plan view)

Turbulent wake

(c) Flow separation at a bend (plan view)

Turbulent wake

Recirculating flow

Figure 6.7 (a) Flow separation around a boulder on the bed of a river. (b) Flow separation around an obstruction. (c) Flow separation at a bend. (b) and (c) adapted from Morisawa (1985).

a fluid is determined by its structure at the molecular level, as work must be done to move the molecules past one another. Factors such as the regularity of molecular shapes and the strength of attraction between molecules affect the way in which the fluid responds to deformation. The more viscous a fluid is, the more it resists deformation and the less easily it flows.[4]

The Reynolds number, Re, is given by:

$$Re = \frac{Inertial\ forces}{Viscous\ forces} = \frac{vR}{\mathbf{v}}$$

where: Re = Reynolds number, v = mean flow velocity, R = hydraulic radius, ρ = fluid density and $\mathbf{v}$ (Greek letter nu, which confusingly looks rather like a 'v') = kinematic viscosity. At low Re numbers (less than 500), the viscous forces dominate and flow is laminar. Where the inertial forces are dominant (at Re numbers greater than 2,100), the inertia of the flowing water is much more significant than the viscous forces resisting that movement and turbulent flow occurs. The transition between laminar and turbulent flow occurs between Re values of 500 and 2,000.

The Reynolds number is dimensionless: it does not have units. When velocity (in m s^{-1}) and hydraulic radius (m) are multiplied together, the resultant units are m^2 s^{-1}. The units for kinematic viscosity are also m^2 s^{-1}, so when the Reynolds number is calculated the units cancel out.

The boundary layer

When a fluid moves over a solid boundary, such as the wall of a pipe or the bed of a channel, it is affected by friction between the fluid and the boundary, in addition to the internal friction (viscosity) within the fluid itself. At a certain distance from the boundary, its effects are no longer 'felt' by the fluid and the flow velocity reaches a maximum or **free stream velocity**. The **boundary layer** is the thickness of flow that is affected by the boundary and is significant for several reasons. Much of the erosion and transport of sediment takes place in the boundary layer, turbulence is generated within it, and most of the plants and animals that are found in rivers live within this zone.

Structure of the boundary layer

A velocity gradient exists through the boundary layer, with the velocity increasing with distance from the boundary, at which it is zero. Different zones can be identified within the boundary layer as indicated in Figure 6.8(a). Immediately above the boundary is a thin (a millimetre or less) **viscous** or **laminar sublayer** within which the fluid movement is slowed so much by friction that the flow is laminar. Despite its limited thickness, the laminar sublayer is not insignificant. Fluid shear stresses within this layer are low and small particles of sediment that are wholly submerged within

it are 'protected' from turbulent eddies that may entrain those particles that project above it. It also provides protection to small organisms. Between the laminar sublayer and the outer, turbulent, boundary layer is a transitional or **buffer layer** in which the flow structure is intermediate between laminar and turbulent. The outer boundary layer is fully turbulent and is called the **logarithmic layer**. Within this layer, the time-averaged velocity is often observed to increase logarithmically with height above the boundary (Richards, 1982). Above the boundary layer, in the **free stream layer**, there is no velocity gradient (Figure 6.8b). The free stream layer is not always present because, in many cases, the boundary layer extends through the whole depth of flow.

Hydraulically rough and smooth surfaces

As well as influencing the overall resistance of a channel, roughness elements, such as grains of sediment, have significant effects on the structure of the boundary layer. Significant here is the height of grains relative to the thickness of the boundary layer. Where the grains of bed sediment are small enough to be totally submerged in the laminar sublayer, the flow is described as being **hydraulically smooth** (Figure 6.9a). In this case the flow over the boundary is the same as it would be if the boundary was totally smooth and no grains were present. A different situation arises when grains are large enough to project through the sublayer. The rate of energy loss is increased as a result of turbulent eddy shedding. Eddies are generated as the flow moves over and around the particles, eddy size increasing with particle size (Leeder, 1999). In this situation the flow is **hydraulically rough** (Figure 6.9b). Where the penetrating grains are of a similar diameter to the thickness of the sublayer the surface is described as **transitional**.

To complicate matters, the thickness of the laminar sublayer decreases with increasing near-bed velocity. This means that some of the 'protected' smaller grains can become exposed at higher flows. The spacing of individual grains also has an effect, because eddy development and flow resistance are reduced when grains are closely spaced. The degree of channel bed roughness

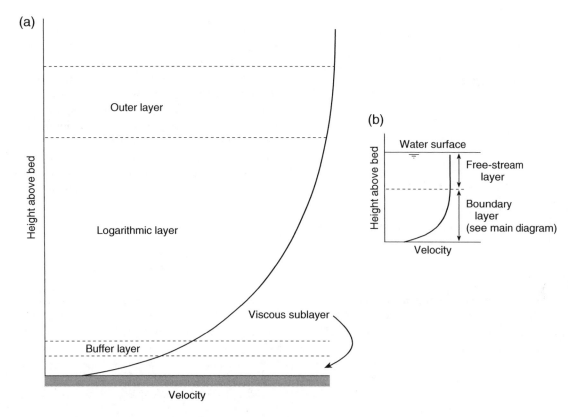

Figure 6.8 (a) Velocity profile and boundary layer for turbulent flow. This shows how the average velocity (horizontal axis) varies with the height above the bed surface. (Note that the thicknesses of the different layers are not to scale.) After Robert (2003). (b) Fully developed flow profile.

can be defined using a **grain Reynolds number (Re*)** (this is also called the **boundary** or **shear Reynolds number**). This is proportional to the ratio between the grain size and thickness of the laminar sublayer and is defined in Box 6.4. At low Re* values, the grains are contained within the laminar sublayer and the surface is hydraulically smooth. As Re* increases (for larger grain sizes, or where the thickness of the laminar sublayer is reduced), grains start to project through the sublayer. The flow is then transitional or rough.

Most natural channels are hydraulically rough (Robert, 2003), with roughness elements including coarse sediment, bedforms and woody debris. The form roughness associated with bedforms can be very high because of the generation of eddies associated with flow separation (Leeder, 1999).

Momentum transfer, velocity distributions and fluid shear stress

The fact that a velocity gradient exists within the boundary layer – or that there is a boundary layer at all – is due to the viscosity of the fluid. If the fluid were non-viscous (something called an 'ideal fluid'), it would all flow at the same velocity, with the exception of a thin layer of molecules that adhered to the boundary itself.

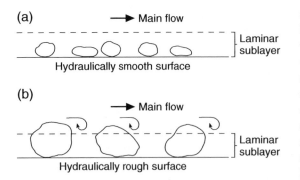

Figure 6.9 (a) Hydraulically smooth surface.
(b) Hydraulically rough surface. Adapted from Morisawa (1985).

However, most fluids, including water, are not 'ideal' and something called **momentum transfer** takes place. The momentum of a moving object, or 'parcel' of fluid, is determined by the product of its mass and velocity. Within the flow, momentum transfer allows slower moving fluid to be speeded up by faster moving fluid,

and vice versa. For laminar flows this is brought about by a process called **molecular diffusion**. Molecules within a slower moving layer of fluid have less momentum than molecules in the faster moving layer above. If a molecule moves from a slower moving layer to a faster moving one, that molecule will have less momentum than the molecules surrounding it. As a result it will be speeded up and, at the same time, it slows down the surrounding molecules very slightly. Similarly, a molecule moving from a faster moving layer to a slower moving layer will be slowed down by the surrounding molecules, but will accelerate them slightly.

Molecular diffusion also occurs within turbulent flows, although it is insignificant in comparison to a much more effective form of momentum transfer brought about by turbulent eddies that transfer 'lumps' of fluid within the flow. In a recognisable sequence of events, called **turbulent bursting**, an **ejection** of low momentum fluid occurs from the region close to the boundary. This moves upwards into the flow profile. Following this, an inrush, or **sweep**, of high momentum fluid moves downwards to replace the ejected fluid.

Box 6.4

GRAIN REYNOLDS NUMBER

The particle Reynolds number (Re*) combines the effects of near-bed velocity and grain size and allows comparison of the size of surface irregularities with the thickness of the laminar sublayer:

$$\mathrm{Re}* = \frac{v* D}{v} = 11.6 D / \delta_{\mathrm{sub}}$$

Where v is kinematic viscosity, D is the grain diameter, $v*$ is the shear velocity and δ_{sub} is the thickness of the laminar sublayer. **Shear velocity** is a measure of shear stress and velocity gradient close to the boundary and has units of velocity (m s^{-1}). A large shear velocity

implies a steep velocity gradient and a large shear stress (Chanson, 1999). Shear velocity is given by:

$$v* = \sqrt{\tau_0 / \rho}$$

where τ_0 is bed shear stress and ρ is fluid density.

For hydraulically smooth flows Re* is less than about 5 (Schlichting, 1979). This means that the particle diameter, D, is less than the thickness of the laminar sublayer, δ_{sub} (remember that the thickness of the laminar sublayer varies with near-bed flow conditions, so some grains can become exposed as the near-bed velocity gradient increases). Hydraulically rough flows are associated with a Re* greater than 70, while transitional flows occur between 5 and 70 (Schlichting, 1979).

Box 6.5

SHEAR STRESS DISTRIBUTION WITHIN THE VELOCITY PROFILE

The velocity gradient can be calculated in a similar way to a topographic gradient. Instead of working out the change in height over a given distance, we are interested in the change in velocity over a vertical distance (i.e. difference in height above the bed). This is done by dividing the difference in velocity between two points (dv) by the vertical difference (dy) between those points, i.e. dv/dy.

Since there is a velocity gradient, there must also be a shear stress between different 'layers' of the fluid, which are travelling at different velocities. This acts over the area of the plane of contact between the two layers. The steeper the velocity gradient between layers, the larger the shear stress between those layers.

In laminar flows, the equation relating shear stress (τ) to the molecular viscosity (μ), and the velocity gradient (dv/dy) is:

$$\tau = \mu \frac{dv}{dy}$$

This equation tells us that the shear stress is greater for viscous fluids and where there is a steep velocity gradient.

The shear stress for turbulent flows is determined, as for laminar flow, from the molecular viscosity and the velocity gradient, with the addition of an extra term, ε (eta), to represent the 'eddy viscosity'.

$$\tau = (\varepsilon + \mu)\frac{dv}{dy}$$

As ε is much, much larger than μ, the latter can be ignored, giving:

$$\tau = \varepsilon \frac{dv}{dy}$$

Unlike molecular viscosity, the value of the eddy viscosity, ε, increases with increasing velocity as a result of increased turbulence intensity. Turbulent flow calculations are further complicated by the fact that turbulence intensity increases with distance away from the boundary.

This sequence of events recurs with a certain periodicity and is strongly related to flow structures called vortices, which develop within the transitional zone of the boundary layer (see Leeder, 1999, Robert, 2003, or Bridge, 2003, for more detail).

Momentum transfer by turbulent eddies results in an apparent viscosity which is called **eddy viscosity**. The large size of eddies means that this type of momentum exchange is much more efficient than molecular diffusion. This means that high momentum fluid and low momentum fluid are much more thoroughly 'mixed' within the turbulent region. As a result, velocity differences between different 'layers' of flow in the turbulent part of the profile are smaller than those observed in the underlying viscous layers (laminar sublayer and buffer zone). In other words, the velocity gradient is much

gentler in the fully turbulent logarithmic layer than in the underlying viscous layers. This can be seen from Figure 6.8(a).

Since there is a velocity gradient, there must also be a shear stress between different layers within the flow, which are travelling at different velocities. This acts over the area of the plane of contact between the two layers. The steeper the velocity gradient between layers, the larger the shear stress between those layers. Box 6.5 provides more detail.

Calculating bed shear stress

Much research has focussed on defining how the time-averaged velocity varies with height above the bed within turbulent boundary layers. This is difficult to predict, mainly because the eddy viscosity varies with the nature of the turbulence. Also the situation is greatly complicated when the boundary is made of movable sediment. Bedforms shaped by the flow modify the geometry of the channel. This, in turn, feeds back to affect the flow.

An understanding of the velocity–height relationship brings with it a number of practical applications. For example, it allows bed shear stress to be derived from measured vertical velocity profiles. Bed shear stress is used in numerous equations to describe flow, to determine boundary resistance and to estimate the volume of sediment that is being transported. Unfortunately it is extremely difficult to measure directly in the lab, let alone in natural river channels (Middleton and Southard, 1984). If the flow is steady and uniform – rarely the case in natural channels – the du Boys equation can be used (Box 6.1). This involves measuring the channel or water surface slope using surveying techniques but is not necessarily a straightforward procedure.

An alternative method is to measure the time-averaged velocity at different heights throughout the turbulent profile. From these measurements, the bed shear stress can be calculated indirectly using a relationship called the logarithmic 'law of the wall'. This describes the relationship between vertical velocity distribution, boundary roughness and a 'surrogate' for bed shear stress called the shear velocity (introduced in Box 6.4). This a measure of the velocity gradient and shear stress near the boundary (Chanson, 1999).

Bed shear stress is related to the near-bed velocity gradient, increasing with the steepness of the gradient. It is also affected by the roughness of the channel bed. The 'law of the wall' is explained in Box 6.6, along with some of the assumptions that are made in applying it.

Secondary flows

Secondary flows are vertical and lateral currents that develop within the channel, perpendicular to the main direction of flow. They exist at a much larger scale than turbulent eddies but are weaker than the primary (downstream) flow. This cross-channel circulation is superimposed on the primary flow to produce helical secondary flow cells. Secondary flows are caused by irregularities in the channel boundary, or where there is a difference in water surface elevation across the channel. This situation occurs at meander bends, where water 'piles up' at the outside of the bend. This sets up a pressure gradient, causing water to move downwards and across the channel bed, from the outside to the inside of the bend. Secondary flows are also observed in straight channels, with the spirals scaled to the depth of flow. Sediment movement associated with secondary flows at meander bends will be discussed in Chapter 8.

Overbank flows

Overbank flows occur when the capacity of the channel is exceeded. As mentioned in Chapter Three (p. 30) there is a certain amount of spatial variation along the channel in the timing and extent of inundation. Rather than a river suddenly 'bursting its banks', inundation first starts to occur in those areas where the bank topography is of a relatively low elevation. During peak flow conditions, the water mainly flows in a down-valley direction (Bridge, 2003).

Overbank flows are more complex than within-channel flows, with the channel and floodplain forming what is called a compound channel. This consists of a deeper, central portion (the river channel) which is flanked on both sides by shallower floodplain flows. A section across the floodplain and near-bank region of a straight, trapezoidal channel during overbank flow conditions is illustrated in Figure 6.10. This is a conceptual model, developed by Knight and Shiono (Knight, 1989; Shiono and Knight, 1991), which represents the interactions between the main channel and floodplain flows. In general, the velocity of flow on the floodplain is much lower than that in the main channel because of the shallower depth of flow, which creates a very large wetted perimeter. In addition, floodplain flows are greatly retarded by the increased roughness of the floodplain surface. This is due to vegetation (especially bushes and scrub), man-made structures such as field boundaries, and variations in floodplain topography. Within the channel itself, experimental and field studies have shown that flow velocities are relatively fast (Bridge, 2003). This is indicated by the local and depth-averaged velocities shown in Figure 6.10. As a result, a shear layer develops between the faster-moving channel flow and slower-moving floodplain flow. This extends

Box 6.6

THE LOGARITHMIC 'LAW OF THE WALL' FOR CALCULATING BED SHEAR STRESS

The logarithmic law of the wall

Within the viscous-dominated layers (laminar sublayer and buffer zone) there is a linear increase in velocity with height above the channel bed. In this zone, the velocity gradient is proportional to the bed shear stress. Above this, in the fully turbulent zone, observations have shown that the time-averaged velocity increases with height above the bed according to:

$$v_y = b \ln (y / y_0)$$

where v_y is the time-averaged velocity at a given height, y above the bed. Two other terms need to be defined: b and y_0.

The term b is a measure of the velocity gradient (Robert, 2003). It can be shown (Richards, 1982; Middleton and Southard, 1984) that:

$$b = 2.5 \sqrt{(\tau_0 / \rho)}$$

where τ_0 is bed shear stress and ρ is fluid density. The term $\sqrt{\tau_0 / \rho}$ is known as the **shear velocity** ($v*$) and is a measure of shear stress and velocity gradient. A large shear velocity implies a steep velocity gradient and a large shear stress (Chanson, 1999). The term b can therefore be rewritten as:

$$b = 2.5 v*$$

The other term, y_0, is related to the roughness of the bed. It is the (imaginary) height above the boundary at which the time-averaged velocity is equal to zero (Richards, 1982).[*] This is determined by the **roughness height**, which relates to the size of bed sediment. In hydraulically smooth flows, y_0 is independent of the roughness height since the bed sediment is 'hidden' within the laminar sublayer. However, a decrease is seen in y_0 as the shear stress increases (Richards, 1982). In rough flows, the height above the boundary

at which the velocity is zero (y_0) is similar to the size of the particles. y_0 can be defined in terms of a characteristic grain size, D_{65} (the size for which 65 per cent of all the bed sediment is finer):

$$y_0 = 0.033 D_{65}$$

This is called the Nikuradse sand roughness and was derived from experiments carried out by the engineer and physicist Johnson Nikuradse using flow through pipes that had been lined with an adhered layer of sand grains.

Substituting the shear velocity term in the first equation above, the law of the wall equation can be re-written as:

$$v_y = 2.5 \; v* \ln (y / y_0)$$

This shows that the mean velocity decreases with increasing roughness (represented by y_0). Also, for a constant mean velocity, higher roughness leads to a lower near-bed velocity. This means that the near-bed velocity gradient is steeper. As a result the bed shear stress is greater.

Using the law of the wall to determine bed shear stress

If measurements of time-averaged velocity are plotted against height above the bed on semi-logarithmic paper, a linear relationship can be expected (Robert, 2003). The bed shear stress can be estimated from the slope, b, of the regression line:

$$\tau_0 = (b / 2.5)^2$$

A full description of how this is done is provided by Robert (2003), who also discusses problems associated with the actual measurement of the velocity profile.

Box 6.6

THE LOGARITHMIC 'LAW OF THE WALL' FOR CALCULATING BED SHEAR STRESS—CONT'D

Assumptions and limitations

Two important assumptions are made in the derivation of the law of the wall. The first is that the near-bed fluid shear stress is approximately equal to the bed shear stress. The second assumption relates to something called the **mixing length**, which is associated with momentum transfer. It can be defined as the characteristic (vertical) distance travelled by a particle of fluid before its momentum is changed to that of the new environment (Chanson, 1999). The mixing length varies according to the height above the boundary. In the law of the wall equation, it is assumed that

this is a linear variation and is represented by the constant (2.5) in the law of the wall equation.

These two assumptions mean that the law of the wall is generally only applicable in the lower 20 per cent of the flow profile (Robert, 2003). It has been shown that measuring velocity for a number of points throughout the full flow depth, rather than the bottom 20 per cent of the profile, leads to an underestimation of the shear stress under uniform flow conditions (Biron *et al.*, 1998). Sophisticated measuring devices are required to make velocity measurements close to the bed.

---

*Although the flow velocity is only equal to 0 at the boundary itself, the term y_0 is actually an artefact of the solution of the equation (Bridge, 2003).

---

laterally for a considerable distance, both across the floodplain and into the main channel. The shear layer indicates a two-way transfer of momentum between the floodplain flow and main channel flow (Knight, 1989). This comes about because the slower moving flow on the floodplain reduces the velocity of the flow in the channel. At the same time, the faster-moving flow in the channel speeds up the flow on the floodplain. Associated with the shear layer is a bank of large-scale vertical 'interface vortices' (shown in Figure 6.10), which transfer high-momentum fluid from the channel onto the floodplain (Knight and Shiono, 1996). This has important implications for sediment movement, since sediment is also routed onto the floodplain.

CHAPTER SUMMARY

The flow of water is driven by the downslope component of gravity. Forces that resist this movement include friction between the flow and channel boundary, and viscous forces within the fluid. Boundary resistance refers to the characteristics of the channel boundary.

It has two components: grain roughness relates to the effects of the individual grains making up the channel boundary, while form roughness is associated with features such as ripples and dunes. At the valley scale, channel resistance is associated with downstream changes in valley alignment and channel form. As water flows down-slope, potential energy is converted to kinetic energy. Most of this energy is used in overcoming flow resistance, when kinetic energy is converted to heat energy as a result of friction. Most of this friction occurs within the moving fluid as a result of turbulence. Heat is dissipated to the surrounding environment, resulting in a downstream reduction in total energy along the channel. The flow in natural channels is unsteady, varying over time in response to fluctuations in discharge. Downstream changes in channel shape mean that the flow is spatially non-uniform. Variations in velocity are seen at different scales, including small-scale turbulent fluctuations, changes with depth, downstream changes and variations associated with unsteady flow.

The balance between inertial and gravitational forces determines whether flow is subcritical, critical,

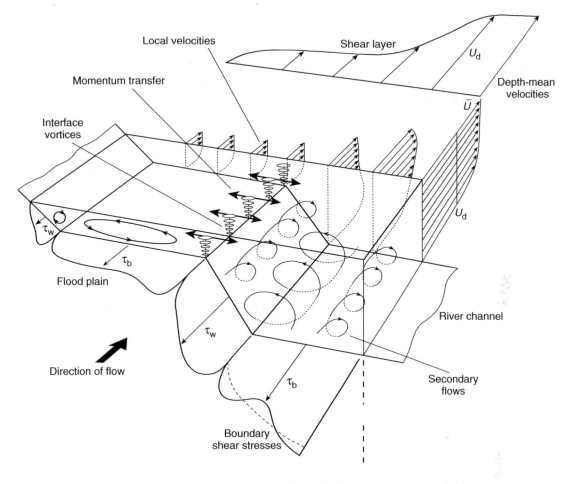

Figure 6.10 Hydraulic parameters associated with overbank flow. This shows a section across the floodplain and near-bank region of a channel during overbank flow conditions. A bank of interface vortices forms at the channel margin as a result of intense shear between the faster channel flow and slower floodplain flow. Fluid is transferred to the floodplain by convection. Local, time-averaged, flow velocities are shown at different verticals; depth mean velocities are also indicated. Also shown is the distribution of boundary shear stress. (τ_0 is shown as τ_b and τ_w referring to shear stress at the bed and shear stress at the banks respectively.) The reduction in shear stress on the floodplain is associated with a reduction in turbulence intensity and sediment-transport capacity (see Chapter 7). After Shiono and Knight (1991).

or supercritical. Flow can also be classified as laminar or turbulent according to the ratio between inertial and viscous forces. At low flow velocities, viscous forces dominate and the flow is laminar. At higher flow velocities, inertial forces dominate and a turbulent flow profile develops. The boundary layer is the depth of flow that is affected by friction with the boundary. A velocity gradient exists through this layer and different zones can be identified within it. Very close to the bed, the viscous layers include the laminar sublayer and buffer layer. Within the outer part of the boundary layer the flow is fully turbulent. A shear stress is exerted at the channel boundary by the flow. This bed shear stress is related to the near-bed velocity gradient. It is also affected by the roughness of the channel boundary. Flows are hydraulically smooth if the bed particles are small enough to be enclosed within the laminar sublayer. Where grains penetrate into the turbulent part of the profile, the flow is hydraulically rough.

FURTHER READING

Introductory texts

Kay, M., 1998. *Practical Hydraulics*. E & FN Spon, London. Written for a wide audience, there are several sections of relevance in this excellent, non-technical introduction to hydraulics.

Advanced texts

Leeder, M., 1999. *Sedimentology and Sedimentary Basins: From Turbulence to Tectonics*. Blackwell, Oxford. Contains two interesting chapters on fluid flow for those with a basic understanding of mathematics.

Middleton, G. and Southard, J.B., 1984. *Mechanics of Sediment Movement*. Society of Economic Palaeontologists and Mineralogists (SEPM). Short Course No. 3. Provides some excellent descriptions and clear explanations for those wishing to develop a rather more advanced understanding, although a knowledge of some calculus and basic Newtonian physics is required to follow it fully.

Richards, K., 1982. *Rivers: Form and Process in Alluvial Channels*. Methuen, London. Long out of print but available in libraries, this classic text provides clear, but quite mathematical, explanations at an advanced level.

Robert, A., 2003. *River Processes: An Introduction to Fluvial Dynamics*. Arnold, London. Provides more detail on stream flow processes, with the emphasis on turbulence and its importance in the study of fluvial processes.

Techniques

Information on recent developments in flow measurement and characterisation is provided by Robert (2003). See also:

Whiting, P., 2003. Flow measurement and characterisation. In: G.M. Kondolf and H. Piégay (eds), *Tools in Fluvial Geomorphology*. John Wiley & Sons, Chichester, pp. 323–46.

7

PROCESSES OF EROSION, TRANSPORT AND DEPOSITION

The three components of a river's load – bedload, suspended load (including wash load) and dissolved load – were introduced in Chapter 5. This discussed the large-scale transfer of sediment through the fluvial system. The actual processes of sediment transport will be considered here, with an emphasis on bedload transport. Interactions between flow and the channel boundary lead to adjustments of channel form, which result from processes of erosion, transport and deposition. In this chapter you will learn about:

- Processes of erosion in bedrock channels.
- Processes of bank erosion in alluvial channels.
- The entrainment (setting in motion) of sediment particles.
- How sediment is transported in river channels.
- How, where and why sediment is deposited.

THE CONCEPT OF STREAM POWER

Sediment entrainment, transport and deposition all involve the interaction of forces. **Work** is carried out when a force moves an object, the amount of work being defined by the size of the force and the distance over which the object moves. Work is involved in moving water through the channel and in eroding and transporting sediment. **Energy** is the capacity or ability to do work, and the same units, joules (J), are used for both.[1] **Power** defines the rate at which work is done and is measured in watts (W), or joules per second. The concept of power can be illustrated by considering the transport of a piece of gravel between two points. This could be accomplished in a short period of time by a large force (high power), or over a longer period by a smaller force (low power). Although the same amount of work is involved in each case, it is carried out at different rates.

Stream power is measured in watts per unit length of stream channel, usually $W\ m^{-1}$. Stream power determines the **capacity** of a given flow to transport sediment. This is the maximum volume of sediment that can be transported past a given point per unit time. The available stream power is related to the water surface slope (S) and discharge (Q) of the channel. It is also affected by the gravitational constant, g, and the mass density of the fluid ($1,000\ kg\ m^2$ for water), which is represented by the Greek letter rho (ρ). These are

---

Box 7.1

EXAMPLE APPLICATION OF THE STREAM POWER EQUATIONS

In order to calculate the stream power per unit length for a reach of river channel you would need to measure the slope of the channel and the discharge. Say the channel slope, S, is 0.01 m m^{-1} and the discharge, Q, is 4 m^{3}s. We know the density of water, ρ (1000 kgm^3), and the gravitational constant, g (9.8 m s^{-2}), so:

$$\text{Stream power, } \Omega \ = \ \rho g Q S$$

$$= \ 1{,}000 \text{ kg m}^{-3} \times 9.8 \text{ m s}^{-2} \times$$
$$4.0 \text{ m}^3 \text{ s}^{-1} \times 0.01 \text{ m m}^{-1}$$

$$= \ 392 \text{ W m}^{-1} \text{ or } 0.39 \text{ kW m}^{-1}$$

This is the rate at which potential energy is converted to kinetic and heat energy along each metre of channel length. To calculate the stream power per unit area, it is necessary to measure channel width, W, and divide stream power by this.

Say channel width, W, is 3.7 m:

$$\text{Specific stream power, } \omega \ = \ \frac{\Omega}{W}$$

$$= \ 105 \text{ W m}^{-2}$$

In order to compare the stream power of two different channels, or different reaches of the same channel, it is important to take measurements when the flow conditions are comparable, as it would be fairly meaningless to compare low flow conditions in one channel with bankfull flow in another. For this reason specific stream powers are usually compared for bankfull flows.

---

combined in the equation below, where stream power is represented by the Greek capital letter omega (Ω):

$$\Omega = \rho g Q S$$

Stream power is often defined in terms of the **specific stream power**, or stream power per unit area of the bed (per m^2). Specific stream power (lower-case omega, ω) is calculated by dividing the stream power per metre length of channel by the width of the channel.

$$\omega = \frac{\Omega}{W}$$

Where W is channel width. This is useful for making comparisons between rivers, or different reaches of the same river, because it reduces the scale effects of large and small channels. For British rivers, the specific stream power ranges from less than 10 W m^2 for lowland channels in parts of the south-east, to 1,000 W m^2 for rivers in the north and west, which drain steep upland areas with high rainfall (Ferguson, 1981).

An example application of the stream power equations is shown in Box 7.1.

Specific stream power can be related to bed shear stress (τ_0) and (cross-sectional) average flow velocity ($\bar{v}$):

$$\omega = \tau_0 \bar{v}$$

This means that the power per unit area of the bed is equal to the product of the average bed shear stress and the average flow velocity. Flow **competence** is the ability of a given flow to entrain sediment of a certain size and increases with bed shear stress.

PROCESSES OF EROSION IN BEDROCK CHANNELS

The morphology of bedrock channels is mainly influenced by processes of erosion because the supply of sediment is often limited. Three types of erosion are significant: block quarrying, abrasion and corrosion. **Block quarrying** is the dominant process (Hancock *et al.*, 1998) and involves the removal of blocks of rock

Plate 7.1 Potholes near the site of a waterfall, Blyde River Canyon, South Africa. These have a diameter of 2–3 m. Photograph by Helen Houghton-Carr.

Plate 7.2 Sculpted forms in the rock bed of an ephemeral channel, Atlas Mountains, Morocco. Photograph by Mike Simms.

from the bed of the channel by drag and lift forces. The size of the quarried blocks can be considerable. Tinkler (1993) reports blocks of sandstone 1.2 m × 1.45 m × 0.11 m and 1.0 m × 0.5 m × 0.05 m being removed from the bed of Twenty Mile Creek, Niagara Peninsula, Ontario, during normal winter flows, when the flow depth was less than 0.4 m.

Before blocks can be entrained by the flow, a certain amount of 'preparation' is required to loosen them. Sub-aerial weathering and other weakening processes play an important role in this. Weakening processes described by Hancock *et al.* (1998) include the bashing of exposed slabs by particles carried in the load and a previously undocumented process termed 'wedging', which leads to the enlargement of cracks in the bedrock substrate. This is thought to occur when small bedload particles are able to enter cracks that are momentarily widened by fluid forces. The particles then become very firmly lodged and prevent the crack from narrowing again. As time progresses, further widening of the crack can be sustained as larger particles fall into it, and may ultimately lead to block detachment. Under conditions of very high flow velocity, sudden changes in pressure can generate

shock waves that weaken the bed by the process of **cavitation**. This effect is caused by the sudden collapse of vapour pockets within the flow (Knighton, 1998).

Abrasion is the process by which the channel boundary is scratched, ground and polished by particles carried in the flow. Erosion is often concentrated where there are weaknesses and irregularities in the rock bed, which allow abrasion to take place at an accelerated rate. This can lead to the development of **potholes**, deep circular scour features that often form in bedrock reaches. Once a pothole starts to develop, the flow is affected, focusing further erosion. Any coarse material that collects in the pothole is swirled around by the flow, deepening and enlarging it, and literally drills down into the channel bed. Over time potholes may coalesce, leading to a lowering of the bed elevation. Plate 7.1 shows how potholes have contributed to bed lowering near the site of a waterfall.

Scouring by finer material carried by the flow, such as sand, leads to the development of **sculpted forms**. These include flutes and ripple-like features, which reflect structures within the flow (Plate 7.2). These are commonly observed on the crests of large boulders and other protrusions into the flow, where flow separation takes place and fine sediment is decoupled from the flow (Hancock *et al.*, 1998). The rock boundary may also be polished by fine material carried in suspension.

Bedrock channels formed in soluble rock are also susceptible to erosion by **corrosion**, especially where the presence of joints and bedding planes allows solutional enlargement. Solutional features such as scallops may also be seen. These spoon-shaped hollows often cover the walls of cave streamways. Their length is related to the formative flow velocity, ranging from a few millimetres (relatively fast flow) to several metres (relatively slow).

Although the actual processes of erosion operate at a small scale, their effects can be seen over scales ranging from millimetres to kilometres. There are several controls on rates of erosion, which influence the processes described above. These include micro-scale (millimetres to centimetres) variations in the rock structure, the larger scale effects of bedding, joints and fractures, and basin-scale influences such as regional geology and base level history (Wohl, 1998).

BANK EROSION IN ALLUVIAL CHANNELS

Processes of bank erosion are important in the development and evolution of different channel forms, while the migration of river channels across their floodplains involves a combination of bank erosion and deposition. Bank erosion can also create management problems when bridges, buildings and roads are undermined or destroyed. Large volumes of sediment can be generated, leading to problems of aggradation further downstream. Land disputes may also arise where boundaries lie along actively migrating river channels.

Rather than being a process in itself, bank erosion is brought about by a number of different processes which can be considered in three groups:

1   *Pre-weakening processes* such as repeated cycles of wetting and drying, which 'prepare' the bank for erosion.
2   *Fluvial processes*, where individual particles and aggregates are removed by direct entrainment.
3   *Processes of mass failure*, which include the collapse, slumping or sliding of bank material into the channel.

Bank material that has been detached remains at the base of the bank until it is broken down *in-situ* or entrained and transported downstream. A balance exists between the rate of sediment accumulation and its rate of removal, which acts as an important control on rates of bank erosion (Carson and Kirkby, 1972). If material accumulates at the base of the bank at a faster rate than it is removed then, to a certain extent, the bank is protected from further erosion. When the opposite situation applies, with bank material being removed faster than it accumulates, bank erosion will continue, sometimes at an increased rate. A third possibility is that rates of supply are the same as rates of removal. The relative rates of accumulation and removal are dependent on the available stream power and the controls on bank erosion discussed below.

Bank materials and weakening processes

The moisture content of the bank is significant, particularly for cohesive bank materials whose strength varies with the level of saturation. A certain amount of water is held in the pores, against the force of gravity, by matric suction forces. These result from surface tension effects, and a negative pore water pressure (less than atmospheric) develops when the soil is not completely saturated. As the soil dries, the strength of the matric suction forces increases as all but the smallest pores are emptied. These forces can be considerable and several authors have observed an increase in the resistance of the bank material to erosion at high matric suctions. However, it has also been suggested that desiccation can lead to higher rates of bank retreat, because the shrinking of clay particles causes cracking and shedding of loose material at the bank surface.

The process of **slaking** occurs when banks are rapidly immersed by floodwaters and air becomes trapped and compressed within the pores. The resultant pressure causes material to become dislodged (Thorne and Osman, 1988). At high flows, banks may become saturated with water from the channel. Saturation also occurs when there is a rise in the water table or during prolonged rainfall. Under these conditions a positive pore water pressure exists between the grains. This weakens the cohesive forces, acting as a lubricant and reducing inter-granular friction.

During cold conditions, the growth of lenses, wedges, and crystals of ice can significantly reduce resistance to erosion, especially where freeze–thaw cycles occur. In temperate regions, the growth of ice needles occurs during moderately sub-zero temperatures. These are elongated crystals of ice that start to grow as the temperature of the air in contact with the bank decreases, growing in the direction of cooling (into the bank). The crystals often lift and incorporate material which then moves downslope or remains as a 'sediment drape' when the ice melts (Lawler, 1988). In colder regions, where rivers freeze over in winter, cantilevers of ice can cause significant damage (Church and Miles, 1982). Where permafrost exists, thermo-erosion niches are cut into frozen banks by the relatively warm water in the channel.

While not a process in itself, the presence of vegetation influences the resistance to bank erosion in various ways. Root networks are particularly important and vegetated banks tend to have a more open structure and be better drained. Vegetation also acts to bind the soil together and increase the shear strength of the bank material. Unlike soil, roots have a very high tensile strength, which means that they are able to resist tension (stretching forces).

Bank erosion by fluvial processes

For any given situation the relative importance of direct entrainment and mass failure is mainly determined by the composition of the bank, although other factors can also be important. Banks composed of sand and coarser particles are non-cohesive and this material is usually detached grain by grain. Although cohesive forces do not exist between the particles, movement is resisted by inter-particle friction and the packing structures holding the grain in place. However, the selective entrainment of finer sands and gravels often leads to a weakening of the overall structure, which may lead to collapse.

In the case of cohesive banks, it tends to be aggregates and crumbs that are detached rather than individual particles. The weakening processes described above are of great importance in assisting fluid forces to detach and entrain aggregates. Once entrained into the main flow, aggregates tend to disintegrate fairly rapidly.

Bank failure mechanisms

Bank failure occurs when bank material becomes unstable and falls or slides to the base of the bank. There are several types of failure, and different failure mechanisms are observed for cohesive and non-cohesive bank materials. Also important are bank height, bank angle, moisture content and the effects of vegetation.

There are some similarities between bank failure mechanisms and the processes of mass wasting discussed in Chapter 4 (pp. 39–42). The stability of banks is determined by the balance between the shear stress exerted by the down-slope component of gravity (driving force) and shear strength of the bank material (resisting force). In cohesive banks, failure occurs across a **failure plane**, the surface within the bank across which shear stress exceeds shear strength. Failure planes can be almost planar (flat) or curved. One of the most common types of failure is illustrated in Figure 7.1(a) and occurs where banks are low, steep and composed of cohesive material. Typically the failure surface is almost planar and vertical, parallel to the bank surface (Plate 7.3, see also Colour Plate 3). Where bank angles are less steep, the failure plane is usually curved and located deep within the bank (Figure 7.1b). Cohesive banks are often most susceptible to failure after a flood wave has passed, when the saturated banks are no longer supported by the pressure of flow in the channel. Non-cohesive banks tend to fail along shallow slip surfaces (Figure 7.1c). Mixed banks are common, typically with fine cohesive sediment overlying non-cohesive material (Plate 7.4). Undercutting of the non-cohesive material by fluvial processes leads to instability of the overlying material. This can cause various types of bank instability, including the cantilever failure illustrated in Figure 7.1(d).

SEDIMENT ENTRAINMENT AND TRANSPORT

The process of particle entrainment

Whether or not a given particle is set in motion depends on the balance between the forces driving and resisting its movement. These are illustrated in Figure 7.2.

Plate 7.3 Slab failures in cohesive banks, River Ure, North Yorkshire, England.

The resisting force is the immersed weight of the particle (in this simple example the effects of neighbouring grains will be ignored). The driving force is provided by the combined effect of two **fluid forces** exerted on the particle by the flow: a drag force and a lift force. The fluid **drag force** acts in the same direction as the flow and can be thought of as the 'force of the flow' that is felt when you wade out into the current of a stream. It comes about because the pressure exerted on an object by the flow is greater on its upstream side than on its more sheltered downstream side. The second fluid force, the **lift force**, acts vertically upwards and is caused by a pressure difference above and below the particle. Water flowing over the particle has to move faster. According to the

Plate 7.4 Mixed bank, where selective entrainment of the underlying non-cohesive material has led to collapse of the cohesive layer above. Gunnerside Gill, North Yorkshire, England.

Bernouilli principle (Box 6.2), an increase in velocity results in a decrease in pressure above the particle, while the pressure below it stays the same. This difference in pressure generates lift. In theory, if this force exceeds the gravitational force, the particle will be lifted from the bed. In practice, the presence of other particles complicates matters considerably.

Sediment transported as bedload is generally gravel-size and larger, although coarse sands may also form part, or all, of the bedload component. Finer bedload, which is too coarse to be transported in suspension, is moved along the bed in a series of short jumps by **salta-tion**. Saltating grains are lifted from the bed at a relatively steep angle by the combined forces of lift and drag. As a grain moves upwards into the flow, the lift force decreases and it starts to fall back towards the bed. The falling grain is carried downstream by the drag

force, following a shallow trajectory towards the bed. Larger particles, which cannot be lifted, are rolled or dragged along the bed. This movement is usually sporadic because of variations in bed shear stress. In addition, particles tend to become lodged behind other particles or obstacles on the bed. The weight of smaller particles carried in suspension is supported by turbulence. Descending saltating grains may also be temporarily lifted upwards by turbulent movements. This is called incipient suspension.

Size-selective theories of sediment transport

A considerable amount of research has focused on deriving critical flow or **entrainment thresholds** from easily measured flow parameters. There are various

(a) Slab failure

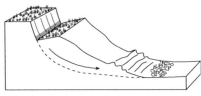

(b) Rotational failure

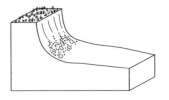

(c) Failure of non-cohesive bank material

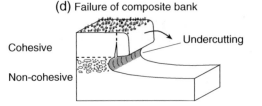

(d) Failure of composite bank

Cohesive

Non-cohesive

Undercutting

Figure 7.1 Bank erosion processes.

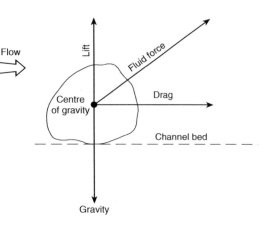

Figure 7.2 Driving and resisting forces acting on a grain of sediment resting on the bed of a channel. Adapted from Morisawa (1985).

practical reasons why we might want to know the flow conditions that will move particles of a certain size. These include the planning of reservoir releases to flush out fine sediment from fish spawning grounds (without removing the gravel), or determining when structures such as bridge piers are at risk of being undermined by erosion.

The threshold conditions for the entrainment of particles of a given size can be defined according to a critical mean flow velocity (i.e. cross-sectional average) or a critical bed shear stress. Using the mean flow velocity is an indirect method, since it is actually the hydraulic conditions near the bed of the channel that are significant. However, both relationships show similar basic trends.

An explanation will first be given in terms of a critical mean flow velocity since this relationship is conceptually easier to understand.

The critical mean flow velocity curves shown in Figure 7.3(a) were derived from a large amount of experimental data accumulated by Filip Hjulstrøm in the 1930s. They show the entrainment and fall (or settling) velocities for particles of different sizes, from fine clay to coarse gravel and small boulders. Note that a logarithmic scale is used on both axes to cover the wide range of particle sizes and the corresponding range of flow velocities. The upper curve on the graph shows the **entrainment velocity** required to set different particle sizes in motion. Sand grains, with a diameter of between 0.2 mm and 0.7 mm, are the easiest to entrain. In the case of larger particles, which have a greater immersed weight, the entrainment velocity increases with particle size as might be expected. However, the relationship is rather different for particles smaller than 0.2 mm, since the entrainment velocity actually *increases* as the particle size decreases from fine sand to silt and clay. Reasons for this include the fact that these small particles tend to be partly or wholly enclosed within the laminar sublayer during most flows. Drag forces are lower within this layer, and particles are not exposed to turbulent lift forces. In addition, the cohesive forces between clay particles further increases the force required to set them in motion.

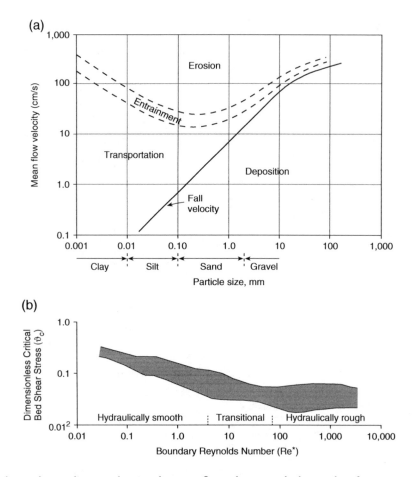

Figure 7.3 (a) The Hjulstrøm diagram, showing the mean flow velocity at which particles of a given size are entrained, transported and deposited. These curves are based on experimental data carried out for well sorted sediments of a uniform size. Adapted from Hjulstrøm (1935). (b) A Shields diagram showing the relationship between the critical dimensionless bed shear stress and the boundary Reynolds number. A large amount of scatter is associated with different experimental conditions and the problem of defining exactly when a threshold has been reached. Adapted from Miller *et al.* (1977) and Summerfield (1991).

An alternative approach, which is more relevant to modern sediment transport theory, was devised by the American engineer Albert Shields in 1936. This defines the critical bed shear stress necessary to set particles of a given size in motion. The critical bed shear stress is actually defined in a dimensionless form. The dimensionless critical bed shear stress is often referred to as the **Shields parameter**. It appears in a number of sediment transport equations and is represented by the Greek letter theta (θ_c – the subscript is short for 'critical'). The equation defining dimensionless bed shear stress is shown in Box 7.2.

Critical bed shear stress increases with particle size but also depends on bed roughness. Shields related

Box 7.2

THE SHIELDS PARAMETER FOR DIMENSIONLESS CRITICAL SHEAR STRESS

The dimensionless critical bed shear stress is defined by:

$$\theta_{cr} = \frac{\tau_{cr}}{g(\rho_s - \rho)D}$$

Where θ_{cr} is the dimensionless shear stress (Shields parameter), g is the acceleration due to gravity, ρ_s and ρ are the density of sediment and water respectively, and D is a characteristic grain size.

the dimensionless bed shear stress to the boundary Reynolds number (Re*). This was defined in Chapter 6 and represents the roughness of the channel bed (see 'hydraulically rough and smooth surfaces' on p. 84–5 and Box 6.4). The boundary Reynolds number is proportional to the ratio between grain size and laminar sublayer thickness. Figure 7.3(b) illustrates the Shields relationship, which is basically similar to the Hjulstrom relationship. Where Re* is less than about 5, the grains are small enough to be fully submerged within the laminar sublayer. As these sheltered particles get smaller, the shear stress needed to entrain them increases. For hydraulically rough surfaces, the critical bed shear stress is independent of the boundary Reynolds number and the critical bed shear stress reaches a constant value of 0.06 (Richards, 1982). The lowest critical bed shear stress is associated with sand grains in the size range 0.2 mm to 0.7 mm (Knighton, 1998).

It is important to note that the Hjulstrom and Shields experiments were carried out using well sorted bed sediment of a single size. This is not representative of the conditions on the bed of many channels, where there is a mixture of grain sizes. The arrangement of grains on the bed and the mixture of grain sizes is very significant, affecting both the entrainment of individual grains and overall transport rates.

Sediment transport in mixed beds

The mobility of individual particles is greatly affected by the size and arrangement of the particles surrounding them. In most natural channels the mixture of sediment sizes, and an irregular bed surface, makes the situation rather more complicated. Figure 7.4(a) shows how the arrangement of grains affects the 'rollability' of individual particles, which determines how easily a particle can be moved from its resting position (Pye, 1994). This can be defined in terms of a **friction angle**, which is greatest where small particles overlie larger ones, meaning that a greater force is required to pivot smaller particles away from the bed. Larger grains can also shelter smaller grains from flows that would otherwise be competent to entrain them (Figure 7.4b). The degree of **sorting** reflects the range of particle sizes in a particular sample of bed material. Well-sorted sediments have a narrow range of particle sizes, whereas poorly sorted material shows a much wider range. Box 7.3 explains how the distribution of grain sizes in a channel are described.

In gravel-bed rivers, particles may also be wedged together in various types of packing arrangements which act to resist bed shear stresses and again make it much harder for individual grains to be entrained. Figure 7.4(c) shows how coarse gravels and pebbles can be deposited in an **imbricated** structure, inclined in the direction of the flow that deposited them, with the long axis of each particle overlapping the next.

Armour layers

In gravel-bed rivers the development of an **armour layer** has a very significant impact on rates of bedload transport. Armouring is illustrated in Figure 7.4(d) and develops during frequent low magnitude flow events that are only competent to entrain the smaller particles. As a result, fine sediment is removed from the bed leaving a layer of coarse sediment, usually about one particle diameter in thickness. This armour layer protects the finer material beneath from subsequent high flows. Once a bed is armoured, a much higher critical threshold is required to break it up. Bathurst (1987a) defined **'two-phase' flow** for armoured channels. During phase 1 flow an armour layer is present and rates of bedload transport

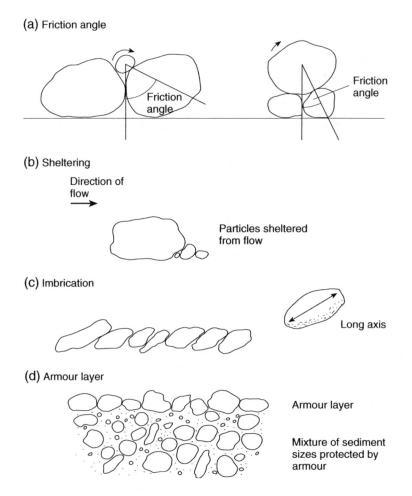

Figure 7.4 Bed sediment characteristics. (a) Friction angle. (b) Sheltering. (c) Imbrication. (d) Armour layer.

are low (although finer sediment can still be supplied from further upstream). Once the armour layer breaks up phase 2 transport takes place, with a dramatic increase in transport rates as the finer sediment becomes available. This can lead to complex variations in bedload transport through time. For example, where two high magnitude flow events occur in close succession, the initial rate of transport is often much higher for the second event than for the first, which breaks up the armour layer. Research has shown that ephemeral channels do not tend to develop an armour layer because, in the absence of low flows, there is no mechanism for removing fine sediment to create the armour layer. This could mean that rates of bedload transport are greater for ephemeral channels than for channels in humid settings (Nanson *et al.*, 2002).

The theory of equal mobility transport

On a level bed with a uniform sediment size, all the particles might be expected to begin moving under approximately the same flow conditions (Reid *et al.*, 1997).

Box 7.3

DESCRIBING THE SIZE DISTRIBUTION OF BED SEDIMENT

With the exception of sand-bed streams, a wide range of different particle sizes is often found within the bed material of river channels. For gravel and boulder-bed streams this can vary over five orders of magnitude from fine clay (< 0.004 mm) to boulders (> 256 mm). Where there is a wide range of sizes it can be difficult to define a representative particle size. The frequency distribution (number of particles in each of a number of specified size classes) of bed material samples is not normally distributed, typically having a positive skew towards the finer particles, which means that there is a 'tail' of coarse sediment. This means that most of the particles have a size that is less than the mean. Because of this, the **median** value is often more representative, and is written as D_{50}. This means 'the size for which 50 per cent of the sampled material is finer'. Another size characteristic that is often used is D_{84} which is the size fraction for which 84 per cent of the sediment is finer.

The typically skewed distribution of particle sizes is similar to a type of distribution called a **log-normal distribution**. The particle size distribution can be transformed into a normal distribution using the logarithmic **phi scale**. Phi units (Φ) are determined from:

$$\Phi = -\log_2 D$$

where D is the grain diameter.

For example, a sand grain with a diameter of 0.02 mm has a value of 1.7 on the phi scale. Further examples are shown in the table. You will notice that the spacing

Table 1 Millimetre and phi scale size ranges for different sediment grains

Description	Size (mm)	Phi scale
Clay	< 0.004	> 8
Silt	0.004 to 0.062	8 to 4
Fine sand	0.062 to 0.25	4 to 2
Medium sand	0.25 to 0.5	2 to 1
Coarse sand	0.5 to 2.0	1 to −1
Gravel	2.0 to 16.0	−1 to −4
Pebbles	16.0 to 64.0	−4 to −6
Cobbles	64 to 256	−6 to −8
Boulders	> 256	< −8

of the different size classes (e.g. clay, sand, gravel) in millimetres varies, with ranges of fractions of a millimetre for the smaller silt and sand size classes compared to tens of centimetres for the larger particles.

The logarithmic phi unit is useful in transforming the frequency distribution into a normal (or near normal) distribution, although this is frequently only a rough approximation (Knighton, 1998). When dealing with a log-normal distribution, the mean value is equivalent to the median of the original distribution of grain sizes, which is why D_{50} is commonly used. D_{84} is one standard deviation above the mean and D_{16} one standard deviation below the mean for a normal distribution. Sorting indexes can also be derived, such as D_{84}/D_{16}. For well sorted sediment (small size range, little variation from the mean) the index has a low value while the value for poorly sorted sediment is higher.

However, on mixed beds, the relative size of a given sediment particle determines its degree of exposure to the flow. As a result, larger particles shelter smaller particles, which then require a higher shear stress for entrainment than would otherwise be the case. In contrast, coarser grains are more easily entrained when surrounded by fine grains. This is because they are relatively more exposed to the forces of entrainment (Andrews, 1983). Particles of an intermediate size are relatively unaffected by the sheltering/hiding effects. Empirically, this 'reference size' has been shown to approximate the median size (D_{50}) (Bathurst, 1987b).

On the basis of field data, Parker *et al.* (1982) introduced a theory of **equal mobility** for channel beds composed of a mixture of sediment sizes. This states that the threshold condition for each size fraction is not dependent on the grain size. In other words, the movement of particles of different sizes can be initiated under similar critical flow conditions. The theory of equal mobility transport therefore challenges the size-selective transport theory of Shields (1936). However, any deviation away from an equal mobility condition represents some degree of size-selective transport. Under conditions of equal mobility, the bedload transport rate could be calculated from a single representative grain diameter such as the median size, D_{50} (Parker et al., 1982).

Equal mobility transport is the subject of some debate, however. Field investigations into the occurrence of equal-mobility transport have mainly been carried out in gravel-bed channels, where the largest grains are cobble-size or smaller (for example Andrews, 1983; Ashworth and Ferguson, 1989). Measurements made in rivers during steady uniform flows have shown that the transport of mixed sediment is only weakly size selective at low shear stresses. At higher shear stresses, sediment transport approaches equal mobility (Parker et al., 1982; Andrews, 1983; Marion and Weirach, 2003). However, observations made over a wider range of flows (e.g. Ashworth and Ferguson, 1989; Wilcock, 1992) have emphasised the size-selective nature of gravel transport. Only during the highest flows does sediment transport approach equal mobility. For example, Wilcock (1992) observed a progressive shift away from unequal to equal mobility transport with increasing shear stress, although equal mobility was not observed until the shear stress was over twice the critical stress required to initiate motion. One of the biggest problems associated with these investigations is obtaining sufficient field data to include a representative range of flow conditions (Reid et al., 1997).

BEDLOAD TRANSPORT

Bedload transport does not necessarily take place all the time, and rates may approach zero during low flows. Even when transport is occurring, it is likely that only part of the bed will be mobile at any one time. Part of the reason for this is the uneven distribution of bed shear stresses, which is directly controlled by variations in turbulent fluctuations. Large differences are observed across small areas of the bed and over short periods of time. Sweep fluid motions, inrushes of high momentum fluid from the outer zone of the boundary layer (Chapter 6, p. 86), are particularly effective at entraining bedload particles (Robert, 2003). The ejection of low momentum fluid away from the bed also allows finer sediment to be lifted up away from the bed and into the turbulent profile, maintaining it in suspension.

The availability of bed sediment has an important influence on overall rates of bedload transport in a given reach of channel, and many bedload-dominated channels are transport limited. This means that transport rates might be lower than expected at a particular flow because of a lack of available sediment. This 'lack' does not necessarily refer to the total volume of bed sediment in a reach, more relevant is the availability of sediment of a certain size or calibre. Thus a flow that is competent to transport only fine gravels will not be able to entrain the larger material in a boulder-bed stream, no matter how abundant this is. The supply of bedload can be especially limited in bedrock channels and the flow capacity often exceeds that required to transport the available load.

Bedforms

Bedforms in sand-bed channels

In sand-bed channels the sand grains can be transported at both high and low flows because of their low entrainment threshold. As a result the bed is easily shaped by flows to form periodic features known as bedforms. These have been intensively studied, both in the laboratory and in natural channels, and a recognisable sequence of bedforms develops in response to changing flow conditions (Figure 7.5). For the purpose of explanation, the starting point is assumed to be a plane bed, something that is rare in natural channels because even the smallest flows start to shape the bed. When water starts to flow over a flat bed the sand grains start to move, individually at first and then in patches, until periodic **ripples** develop, with crests perpendicular to the direction of flow. Field and laboratory research suggests that the wavelength, or spacing, of ripples is mainly dependent on particle size and is typically between 150 mm and 450 mm.

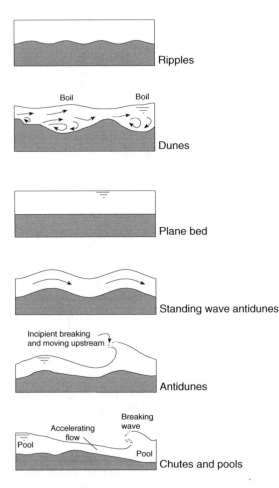

Figure 7.5 Bedform variation in sand-bed channels. Adapted from Simons and Richardson (1966).

where sediment falls down the steeper downstream slope. A critical mechanism in this process is the deposition of coarse grains at the crest, where flow separation occurs.

At higher flows, dunes become unstable and are 'washed out' because the flow velocity is too great to sustain deposition at the dune crest. Dunes then give way to a **plane bed**, but one that is rather different from the initial flat bed. Above the bed is a clearly defined zone of suspended sediment within which 'dust storm conditions' prevail (Leopold *et al.*, 1964). This marks the transition to the **upper flow regime**, where the Froude number (ratio of inertial and gravitational forces) is greater than 1 and flow becomes supercritical. Upper flow regime bedforms include **standing wave antidunes**, where the sediment is moving but the waves themselves are stationary. This is because rates of deposition on the upstream side are matched by erosion on the downstream side. The position of standing waves is marked by waves at the water surface, the sand and water waves being in phase with each other.

At higher flows sediment is thrown up from the downstream side of the bedforms at a faster rate than it can be replenished, which results in **antidunes**. These migrate *up*stream, while the sediment continues to move downstream. Antidunes can be seen in the high-velocity gully flow shown in Colour Plate 15. At very high flows, a series of **chutes and pools** develop. Chutes have a near-plane bed and shooting flow, which enters downstream pools: deeper sections that are marked by hydraulic jumps.

Bedforms in gravel and mixed sand–gravel channels

Bed structures also form in gravel-bed channels and have been a focus of research over recent decades. **Pebble clusters** are commonly found in this type of channel (Figure 7.6) and form when a single large particle acts as an obstacle, protruding into the flow and encouraging the accumulation of coarse material on its upstream side. This upstream material may have an imbricated structure, increasing stability and requiring larger lift and drag forces to entrain the constituent particles. Finer particles are found on the downstream side of the obstacle, where shelter is provided from lift and drag forces. **Transverse ribs** are another type of gravel

As the flow intensity increases, ripples start to give way to **dunes**, larger features with rounded crests. Dunes are common in alluvial channels and are continuous along the bed for hundreds of kilometres in large rivers like the Mississippi and Niger. Dunes vary greatly in size, being scaled with the depth of flow and ranging from a few centimetres to a few metres in height. Dune wavelengths also vary, from tens of centimetres to more than a hundred metres in the largest rivers. Ripples and dunes migrate downstream over time, as the flow moves sand grains up the more gentle upstream slope towards the crest, from

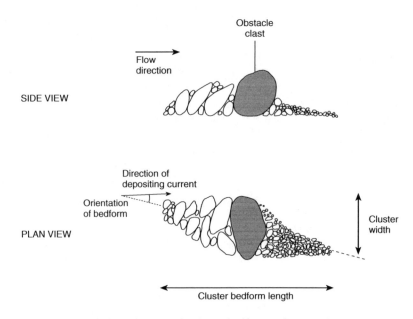

Figure 7.6 Idealised diagram of a cluster bedform. After Brayshaw (1984).

bedform and consist of regularly spaced ridges of coarser pebbles, cobbles or boulders that lie transverse to the flow. Like sand bedforms, these features affect flow resistance as well as rates of bedload transport.

Where the bed is composed of a mixture of sand and gravel, the different mobility of the constituent particles can lead to some interesting effects. For example, longitudinal ribbons of sand have been observed to travel downstream, snaking from side to side over immobile gravel beds. Bedload can also move in thin sheets as an elongated procession of sediment with a thickness of one to two grain diameters. The coarsest sediment accumulates at the leading edge and there is a progressive fining of sediment behind it. Bedload sheets appear to be fairly common in mixed channels and are related to rates of sediment supply, becoming less frequent and reduced in extent as supply rates are reduced (Dietrich *et al.*, 1989).

Assessing rates of bedload transport

From the preceding discussion you will have some idea of just how complex and variable bedload transport is. There is a general paucity of data on rates of transport because the available techniques can be expensive and time-consuming to employ. These include the collection of bedload over a period of time using portable samplers or traps excavated in the bed. Another approach is to track the movement of individual particles. Further information on these techniques can be found in Box 7.4. A number of bedload transport formulae have also been developed, using field and laboratory data, to predict rates of transport from flow variables. An overview is provided in Box 7.5.

SUSPENDED LOAD TRANSPORT

Processes

Particles carried in suspension are kept aloft by turbulent eddies and will remain in suspension as long as their weight is supported by the upward component of turbulent eddies. In a fluid at rest, a suspended particle will fall through the fluid column. The rate of fall, or **fall velocity**, is a function of the density, size and shape of the particle. It is also determined by the viscosity and density of the transporting fluid. Since the falling

Colour Plate 1 Dry ephemeral channel, Eastern Cape, South Africa. The object in the foreground is an old bridge pier.

Colour Plate 2 Anabranching bedrock channel with rock bars, Mpumalanga Province, South Africa.

Colour Plate 3 Ephemeral mixed channel with rock bed and alluvial banks, Eastern Cape, South Africa.

Colour Plate 4 Meandering sand-bed river, the Red River, Texas, United States. Photograph by Airphoto, Jim Wark.

Colour Plate 5 Incised meanders, Colorado River, Arizona, United States. Photograph by Airphoto, Jim Wark.

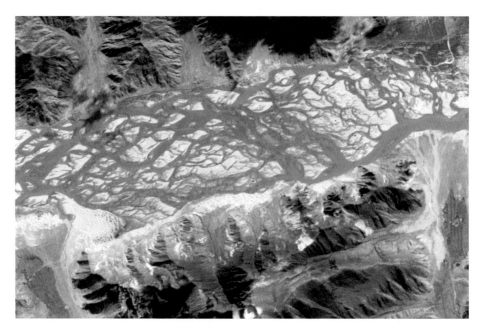

Colour Plate 6 A 15 km stretch of the braided Tsangpo (Brahmaputra) River, near Lhasa, Tibet. Image taken from the International Space Station, courtesy of NASA, Houston TX.

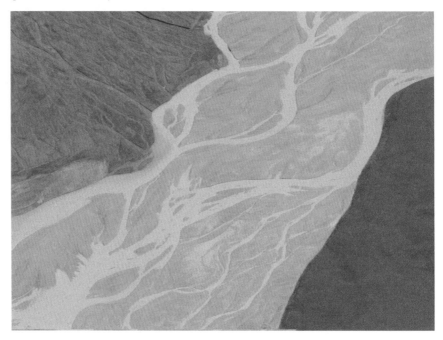

Colour Plate 7 Braided river during low stage conditions, Tien Shan Mountains, Kyrgyzstan. The channel is approximately 500 m wide. A smaller braided tributary joins from the left. Note how the braided pattern is scaled to the size of the two rivers.

Colour Plate 8 Oblique aerial view of Cooper Creek, Queensland, Australia, viewed northwards towards Durham Downs. Photograph by Gerald Nanson.

Colour Plate 9 Columbia River, near Harrowgate, British Columbia, Canada, during the annual summer floods. Levees can be seen as raised ridges bordering the channels and they have been breached in places, resulting in the deposition of splays. Photograph by Airphoto, Jim Wark.

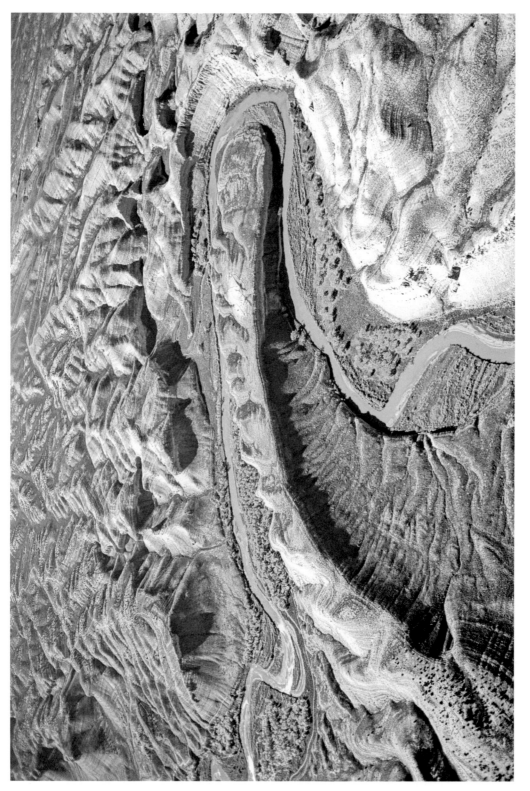

Colour Plate 10 White River, Colorado, United States, an example of a semi-confined valley setting. Meander scrolls can be seen on the inside of bends. Photograph by Airphoto, Jim Wark.

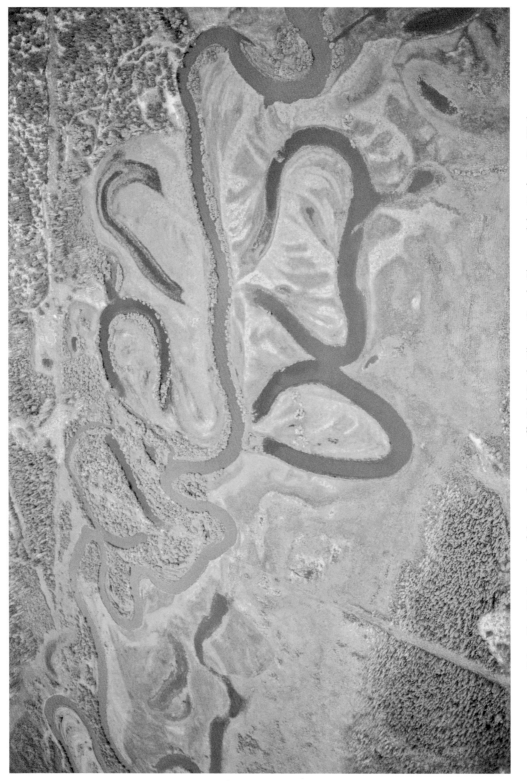

Colour Plate 11 Examples of chute and neck cut-offs. Owl Creek, Alberta, Canada. Photograph by Airphoto, Jim Wark.

Colour Plate 12 The Ganges (flowing from centre left) and Brahmaputra (the large river in the centre of the image) rivers, which both rise in the Himalayas. The Brahmaputra flows west to east across the Tibetan Plateau (top of image), crossing the Himalayas to enter north-eastern India, and Bangladesh. Huge amounts of sediment are generated in the mountains, accumulating on vast floodplains and forming the Ganges delta. The brown area at the Mouths of the Ganges is not land but suspended sediment being discharged to the ocean. The width illustrated is approximately 800 km. MODIS image courtesy of NASA, Houston TX.

Colour Plate 13 The undulating walls of Antelope Canyon, Arizona, United States.
Photograph by Tony Waltham Geophotos.

Colour Plate 14 Flash flood in a normally dry channel, Eritrea. Note the very high concentration of suspended sediment. Photograph by Connell Foley.

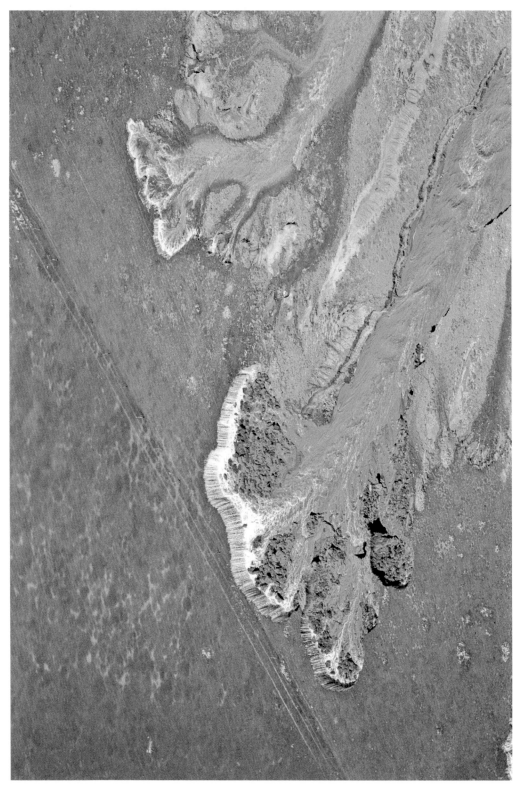

Colour Plate 15 A dramatic example of gully erosion in an arroyo during flood conditions (the green area is totally saturated), eastern Colorado, United States. Photograph by Airphoto, Jim Wark.

Colour Plate 16 Severe aggradation on the Ok Tedi River (a tributary of the Fly River) in Papua New Guinea, caused by large volumes of sediment from gold mining operations. Photograph by Klaus Baumgardt, Foerderkreis 'Rettet die Elbe', Hamburg, Germany.

Colour Plate 17 Wandering gravel bed river, North Slope, Alaska, United States. Photograph by Airphoto, Jim Wark.

Colour Plate 18 Palaeochannels are clearly visible along the valley of the Sweetwater River, Wyoming, United States. Photograph by Airphoto, Jim Wark.

Colour Plate 19 Unpaired terrace sequence formed in glacial outwash deposits, Tien Shan Mountains, Kyrgyzstan.

Box 7.4

FIELD TECHNIQUES FOR ASSESSING RATES OF BEDLOAD TRANSPORT

Portable samplers are widely used. Figure 1 shows one of the most popular devices, the Helley–Smith bedload sampler (Helley and Smith, 1971). This is lowered to the bed of the channel with the entrance facing upstream, and bedload particles entering the sampler are collected in a mesh bag. The length of time over which a sample is collected is dependent on local flow conditions and the volume and calibre of the sediment being transported. The Helley–Smith sampler is designed so that the velocity in the mouth of the sampler is the same as the velocity of the surrounding flow. A problem with earlier samplers was that they tended to disrupt the flow, affecting local rates of transport and reducing the efficiency of the estimates. Despite this, the efficiency of the Helley–Smith sampler does vary according to the size of sediment, being close to 100 per cent for particles ranging from 0.5 to 16.0 mm but decreasing to about 70 per cent for larger material (Emmett, 1980).

Since only part of the bed area is sampled, and for a limited period of time, it is necessary to collect several samples at different locations across the channel, and repeat the whole process at different flows, in order to assess the 'average' bedload transport rate with any degree of confidence. Rates of bedload transport are expressed as the dry weight of sediment passing a cross-section of the river in a given time, with units of kilograms per second. The assumption is usually made that the flow is steady for the duration of each sampling period, although this is not always the case, particularly for flashy mountain streams. Bedforms can also present problems because rates of sediment transport through a particular cross-section vary periodically, reaching a maximum for dune crests and a minimum for troughs. This should be taken into account when deciding on the time interval at which samples are taken.

Bedload traps are a more efficient way of measuring bedload transport, although they involve excavation of the channel and are not practical for large channels.

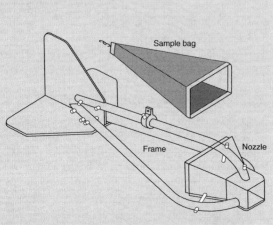

Figure 1 Helley–Smith sampler. After Emmett (1980).

They consist of a trench in the channel, the upstream lip of which is flush with the channel bed, into which any moving bedload falls. The amount of sediment in the trap can be determined by carrying out repeated surveys of the surface, or by removing the sediment and weighing it. There have been various ingenious methods for providing a continuous record, such as the use of pressure sensors to monitor the increase in weight as sediment is trapped over time.

There are alternatives to collecting bedload, one of which is to use tracer particles to determine how far particles of different sizes are transported in a given time. A simple method is to use painted particles. Alternatively, particles can be magnetised and their movement detected using electromagnetic sensors; radioactive tracers are also used. Advances in surveying techniques, including GPS surveying and laser altimetry, mean that repeat surveys of bed elevation can be used to estimate how much bedload movement there has been over a given time. The assumption is that the sediment input into a reach is equal to the sediment output from that reach, with any changes in storage being attributable to variations in bedload movement.

Box 7.5

BEDLOAD TRANSPORT FORMULAE

Most bedload formulae aim to determine the rate of bedload transport as a function of the transport capacity of the flow. This is done by calculating the excess flow capacity above a critical threshold, at which transport is initiated. Below this threshold it is assumed that there will be no bedload transport, above it the rate of transport (volume of sediment passing a given cross-section per unit of time) increases with increasing flow capacity. Many of these formulae have been developed using data for sand-bed channels and should not be applied beyond the range of conditions for which they were formulated.

Flow capacity can be represented using the specific stream power (power per unit width of channel) or average bed shear stress. Many equations are based on the relationship below, which states that the rate of bedload transport, I_b, is a function of the excess stream power required to move the bed sediment.

$$I_b = f(\omega - \omega_c)$$

If the stream power (ω) is less than the critical value (ω_c), no transport will occur. Once the threshold is exceeded, transport will take place, the rate increasing with increasing stream power. One of the more successful bedload equations is that of Bagnold (1980) which was found to perform reasonably well in a comparative study of different bedload equations carried out by Gomez and Church (1989). Bagnold relates the transport rate to the excess stream power and the relative roughness of the channel (ratio of depth to particle size).

Another group of equations uses an excess in bed shear stress:

$$I_b = f(\tau_0 - \tau_c)$$

Where τ_0 is the average bed shear stress and τ_c the critical shear stress. One of the best known examples of this type of formula is the widely used Meyer-Peter and Muller (1948) equation.

When observed rates of bedload transport are plotted against excess shear stress or stream power, the rate of increase is usually non-linear. These relationships are therefore best represented by a power function, giving equations of the form:

$$I_b = f(\omega - \omega_c)^b$$

$$I_b = f(\tau_0 - \tau_c)^b$$

Where the value of the exponent b is usually in the order of 3/2 (Robert, 2003).

A wide variety of bedload formulae have been developed for sand-bed, mixed-size gravel and sand–gravel mixtures. In the case of mixed beds, separate calculations can be carried out for the different size fractions. Empirical hiding functions have also been derived to adjust the relative mobility of each size fraction.

Because bedload transport is so complex, there are several problems and limitations associated with bedload formulae. Many different factors are involved, including flow hydraulics, fluid properties (e.g. viscosity, density and suspended sediment concentration), and the size and arrangement of bed sediment. These interactions are very difficult to represent and there is no general agreement as to which parameters should be included in bedload formulae (Knighton, 1998). Where the supply of sediment is limited, for example as a result of bed armouring, bedload formulae may over-predict actual rates of transport.

A major problem is the limited availability of reliable field data sets for testing bedload formulae. Bedload transport rates are notoriously difficult to measure (see Box 7.4), so the development of bedload formulae has relied heavily on experimental work using laboratory flumes. All bedload formulae include empirical elements, and none is universally applicable to all situations (Reid *et al.*, 1997). This means that each formula should be applied only under conditions that are similar to those for which it was developed.

particle displaces fluid, its movement is resisted by an equal and opposite fluid drag force. If sufficient depth is available, the falling particle will accelerate until it reaches a terminal velocity (defined for falling raindrops in Chapter 4). In channels, the fall velocity is further affected by flow turbulence and the interactions of surrounding particles (Chanson, 1999). Considerable variation is seen between particles of different sizes. The fall velocity for the finest wash load component is very low, meaning that this sediment can be transported over considerable distances (Chapter 5). For example the terminal fall velocity of a silt grain (0.001 mm) is approximately 0.004 cm s^{-1}, but increases to 34 cm s^{-1} for a 10 mm gravel particle (Chanson, 1999).

Suspended sediment is transported by processes of **advection** and **turbulent diffusion**. Advection is the transport of sediment within the flow, where the sediment moves with the flow itself. Turbulent diffusion refers to the mixing of sediment through the depth profile by turbulent eddies. Within the depth profile, the greatest concentration of suspended sediment is found towards the bed of the channel. Although there is continuous movement of individual suspended grains, the overall concentration and average grain size generally decrease rapidly away from the bed. This is due to interaction between the fall velocity and the vertical component of flow associated with turbulent eddying (Knighton, 1998). The upward migration of sediment to zones of lower concentration is both an advective and a diffusion process.

A related process, which is called **convection**, involves the entrainment of sediment by large-scale vortices. For example, sediment is suspended in vortices generated as a result of flow separation in the toughs of ripples and dunes (Bridge, 2003). Large-scale vortices also occur where there are sudden drops in bed elevation, at hydraulic jumps and during overbank flows. Interactions between channel and floodplain flows were discussed in Chapter 6 (see Figure 6.10). Vortices created within the shear zone between the faster moving channel flow and slower flow on the floodplain result in lateral transfers of water. From Figure 6.10 you will see that there is a near-surface flow out onto the floodplain and a return flow back towards the channel. Sediment carried in the flow is also transferred from channel to floodplain. Most of the coarser sediment is deposited near to the channel margins because of the rapid deceleration of flow and reduced turbulence (Bridge, 2003). This explains the origin of alluvial ridges, or natural levees, that are found along the margins of some channels (see Chapter 8). Finer suspended sediment, especially the wash load component, is carried out onto the floodplain.

Sediment supply and transport rates

The main sources of suspended sediment include material washed in from hillslope erosion and the release of fine material and aggregates from bank erosion. The supply of fine sediment is a major control on rates of suspended sediment transport. Most suspended transport, particularly the wash load, is supply limited (Chapter 5, p. 60). This means that the supply of fine sediment often has a greater influence on the sediment concentration than flow conditions in the channel. The rate of supply varies during individual events, between events, seasonally and annually. These variations are controlled by a number of variables, including antecedent conditions, rainfall intensity, hydrograph shape and vegetation growth. High discharges tend to be associated with greater concentrations of suspended sediment. This is because the supply is increased by storm-induced erosion of hillslopes and channel banks, and the release of fine sediment from storage.

With all this in mind, it is hardly surprising that no simple relationship exists between suspended sediment concentration and flow discharge for a given cross-section. This is illustrated by the wide scatter of points in Figure 7.7(a), which shows this relationship for the River Creedy in Devon, England. Figure 7.7(b) shows flow and sediment hydrographs for the River Creedy. For the three events in January 1974 there is a decrease in the sediment load, even though the magnitude of the flow peak of the first two events is similar and the third event has a larger peak flow. This is an example of the 'exhaustion effect' that can result when several events occur in close succession and the available sediment becomes depleted. The role of storage is important here because much of the sediment washed into river channels will have been released from temporary storage on hillslopes.

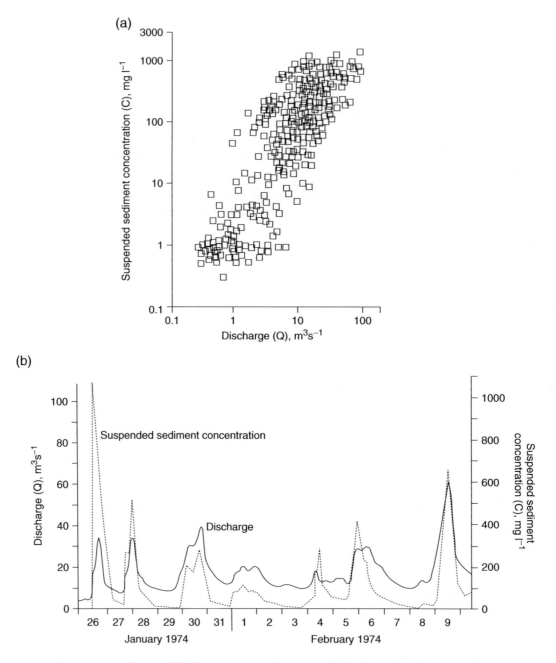

Figure 7.7 Characteristics of suspended sediment transport by rivers. (a) Plot of suspended sediment concentration versus water discharge for the River Creedy, Devon, England. (b) Typical record of the variation of suspended sediment concentration during a sequence of storm runoff events for the same river. After Reid *et al.* (1997).

DEPOSITION

Sediment particles are deposited when there is a reduction in the competence and capacity of the flow. The process itself takes place at a very small scale and involves individual grains, although depositional forms can be observed over a wide range of spatial scales, from the smallest bedforms to vast floodplains and deltas. The construction and development of depositional forms might be likened to the building of an anthill. The process of building the anthill involves individual ants carrying soil one crumb at a time to the site of the ant hill. Although this process takes place at a small scale, the resulting feature is much bigger than the individual ants and crumbs of soil that created it.

Thresholds for deposition are associated with the fall (or settling) velocity defined earlier. The deposition of suspended sediment takes place when the fall velocity dominates over turbulent diffusion. Since the fall velocity is closely related to particle size, the coarsest sediment tends to be deposited first. This leads to sediment sorting, a vertical and horizontal gradation of sediment, from coarse to fine. It should be noted that the fall velocity is also affected by the viscosity and density of the fluid. These are both influenced by changes in suspended sediment concentration. In addition, finer material can be transported as agglomerations of sediment called flocs. These have a greater fall velocity than the individual particles forming them.

In the case of bedload transport, the near-bed flow conditions are significant. Bedload deposition occurs where the bed shear stress drops below the critical shear stress (Shields's parameter) required to transport particles of a given size. Local patterns of sediment sorting are well known, for example a downstream reduction in bed particle size is commonly observed along channel bars (e.g. Bluck, 1982; Smith, 1974).

Where sediment is deposited

There are a number of different circumstances that lead to deposition. These include:

- *Reductions in flow discharge* which are seen as flows recede, or along dryland rivers, where downstream losses are caused by high rates of evaporation and percolation.
- *Decreases in slope* which can be localised, or involve a gradual reduction over a longer length of channel and cause a reduction in average flow velocity and stream power.
- *Increases in cross-sectional area* cause the flow to diverge and become less concentrated. Flow resistance increases because there is more contact between the flow and channel boundary. There is a large increase in cross-sectional area when overbank flows occur.
- *Increases in boundary resistance* are associated with vegetation and coarse bed sediment. When overbank flows occur, velocity is reduced by the increased roughness of the floodplain surface, leading to the deposition of suspended sediment.
- *Flow separation,* which causes sediment to become decoupled from the flow.
- *Obstructions to flow.* Sediment often accumulates behind obstructions. These include boulders, outcrops or islands of bedrock, woody debris and man made structures such as bridge piers, dams and flow control structures.

Changes in the supply of sediment are also important. For example, sediment tends to accumulate immediately downstream from scour zones caused by flow convergence, when the material scoured from the channel bed is deposited immediately downstream. At a larger scale, increases in the supply of sediment to a channel reach are caused by changes within the upstream drainage area (Chapter 5).

Depositional environments

Although deposition does occur in the production and transfer zones of the fluvial system (see Figure 2.1) it dominates in the deposition zone, where there is a decline in gradient and energy availability. Large-scale deposition leads to the development of characteristic landforms, including floodplains, alluvial fans and deltas. Within channels, **bars** represent smaller-scale depositional features. They are commonly found on the inside of meander bends, along the edges of channels, and where tributaries join the main channel. Braided channels are characterised by numerous mid-channel bars.

Floodplains border the channels of alluvial rivers and are formed from a mixture of in-channel and over-bank deposits. Their development and evolution, which will be examined in Chapter 8, is governed by a number of factors, including the supply of sediment (volume and calibre), the energy environment of the channel, and the valley setting. Sediment is laid down by rivers as they migrate across the floodplain, being deposited on the inside of meander bends or when braid bars are abandoned. These channel deposits are relatively coarse in comparison with the much finer sediment that is laid down by overbank flows. Processes of erosion can also be significant in reworking sediment or in removing part, or all, of the flood-plain surface.

Alluvial fans are typically found in situations where an upland drainage basin flows out onto a wide plain (Plate 7.5). The sudden change from confined to unconfined conditions leads to flow divergence, while mean flow velocity is decreased by the reduction in slope. The resultant deposition leads to the formation of a conical feature with a convex cross-profile. Most fans have a radius of less than 8 km, but can be more than 100 km wide in some cases. Where a number of individual fans develop along a mountain front, they may grow laterally and coalesce to form a sloping apron of sediment called a **bajada**. Fans are commonly found in dry mountain regions, where an abundant sediment supply is associated with extreme discharges and fre-quent mass movements. Frequent shifts are often seen in the position of the braided channels that cross the fan surface, although only part of the fan surface may be active during a major flood event. In long profile, the slope is steepest at the fan head, progressively decreasing along the length of the fan. There is also a down-slope reduction in sediment size, although deposits are coarse and poorly sorted. Incision and fan head trenching is associated with decreases in sediment supply, or increases in slope. Such changes can be caused by tectonics, climatic variations, a fall in regional or local base level, or human activity. In the absence of external change, the progressive lowering of the landscape will also result in a decline in sediment yield over time.

Arid fans are generally smaller and steeper than those found in humid regions, a large-scale humid example being the Kosi Fan on the southern Himalayan moun-tain front. This covers an area of 15,000 km^2 and formed where the Kosi River descends onto the wide alluvial plain of the Indus. It has a very low gradient, only averaging 1 m km^{-1} at its head, with further decreases downstream (Summerfield, 1991).

Deltas are found where sediment-charged flowing water enters a body of still water. They extend outwards from shorelines where rivers enter lakes, inland seas and oceans. In coastal areas deltas form where the supply of sediment is greater than the rate of marine erosion, although sediment is redistributed by coastal processes. The influence of fluvial processes tends to dominate in the case of lake deltas.

CHAPTER SUMMARY

Work is involved in transporting water and sediment through the channel. The rate at which work is carried out is defined by the stream power, which is controlled by channel slope and flow discharge. Material is eroded from bedrock reaches by various processes, including block quarrying, abrasion and corrosion. The erosion of alluvial banks also provides a significant proportion of a river's load and is carried out by a com-bination of pre-weakening mechanisms, fluvial action and mass wasting. Whether or not a given flow is able to entrain a stationary particle of sediment on the bed depends on the balance between driving fluid forces (lift and drag), and the resisting force imposed by the particle's submerged weight. The effects of neighbour-ing particles are also important and include inter-particle friction, sheltering, packing and armouring. Rates of bedload transport are strongly influenced by bed sediment characteristics and shear stress distributions, which are highly variable over space and time. Rates of bedload transport can therefore be very difficult to measure or predict accurately. Rates of suspended load transport are influenced mainly by the supply of fine sediment, which is controlled by a number of factors,

Plate 7.5 Alluvial fan descending the northern slope of Adventdalen, a large glacial trough on the main Spitsbergen island, Norway. Photograph by Tony Waltham, Geophotos.

including climate. Suspended sediment is transported by processes of advection and turbulent diffusion. A related process, convection, refers to suspended sediment entrainment by large-scale vortices. Bedload is deposited when the local bed shear stress falls below the critical shear stress for a given particle. The deposition of suspended load occurs when the fall velocity dominates over turbulent diffusion. The fall velocity of a given particle is determined by its density, size and shape, and by the density and viscosity of the fluid. As a result, finer material tends to be deposited first, leading to sediment sorting. Various circumstances lead to sediment deposition, including reductions in slope and discharge, increases in flow resistance, and the divergence of flow around obstructions. Depositional

landforms range in size from small bedforms to vast floodplains.

FURTHER READING

Introductory texts

Reid, I., Bathurst, J.C., Carling, P.A., Walling, D.E. and Webb, B.W., 1997. Sediment erosion, transport and deposition. In: C.R. Thorne, R.D. Hey and M.D. Newson (eds), *Applied Fluvial Geomorphology for River Engineering and Management*. John Wiley & Sons Chichester, pp. 95–135. Provides more detail without getting too technical. Good on monitoring rates of sediment transport.

Advanced texts

Leeder (1999), Middleton and Southard (1984), Richards (1982) and Robert (2003) are all recommended (see 'Further reading' at the end of Chapter 6).

Bridge, J.S., 2003 *Rivers and Floodplains: Forms Process, and Sedimentary Record.* Blackwell, Oxford. Provides a detailed account of sediment transport processes.

Tinkler, K.J. and Wohl, E.E. (eds), 1998. *Rivers over Rock: Fluvial Processes in Bedrock Channels.* Geophysical Monograph Series. American Geophysical Union, Washington DC. This accessible research monograph contains several chapters that refer to erosion processes in bedrock channels.

Classics

Bagnold, R.A., 1966. *An Approach to the Sediment Transport Problem from General Physics.* Professional Paper No. 422–I, United States Geological Survey.

Provides an interesting and readable account of sediment transport processes.

Techniques

Kondolf, G.M. and Piégay, H. (eds), 2003. *Tools in Fluvial Geomorphology.* John Wiley & Sons, Chichester. Includes chapters on bed sediment measurement (Kondolf *et al.*), sediment tracing techniques (Hassan and Ergenzinger) and sediment transport (Hicks and Gomez).

Websites

Sediment transport movies (Department of Geology and Geophysics, University of Wyoming), http://faculty.gg.uwyo.edu/heller/sed_video_downloads.htm. You can download a number of excellent short movies from this site. They include gravel bedload transport, migration of sand bedforms in a laboratory flume, a flash flood and more.

8

CHANNEL FORM AND BEHAVIOUR

The form of a channel is largely a function of the water and sediment supplied to it. Adjustments to channel form occur as a result of process feedbacks that exist between channel form, flow and sediment transport. At the reach scale, the type of adjustment that can take place is constrained by the valley setting, the nature of bed and bank materials, and bank vegetation. This gives rise to a wide diversity of different channel forms. In this chapter you will learn about:

- How the supply of water and sediment drive channel form adjustments.
- Feedbacks between flow, channel form and sediment transport.
- Types of channel adjustment that can be made within the boundaries imposed by local conditions.
- Space and time scales over which adjustments are made.
- Straight, meandering, braided and anabranching alluvial channels.
- Links between channel behaviour and floodplain morphology.
- The form and behaviour of bedrock channels.

CONTROLS ON CHANNEL ADJUSTMENT AND FORM

Flow and sediment supply both fluctuate through time, meaning that continuous adjustment takes place

through the erosion, reworking and deposition of sediment. The flow and sediment regimes are called **driving variables** because they drive these processes.

Along a given reach, channel adjustment is constrained within certain boundaries that are imposed by local conditions. For example, a sand-bed river flowing across a wide floodplain is able to adjust its form much more readily than a bedrock channel confined within a narrow gorge. Energy availability is also important, and channel adjustments are often limited for rivers that flow over low gradients, especially where cohesive banks are protected by vegetation. These constraints are called **boundary conditions** and include valley confinement, channel substrate, valley slope and riparian vegetation. Figure 8.1 provides a schematic representation of the driving variables and boundary conditions that influence channel form.

As discussed in Chapter 3, p. 32 a channel is said to be 'in regime' when its form fluctuates around an equilibrium condition over the time scale considered. Not all channels are in regime, and there are many examples of non-regime, or disequilibrium, channels. This may be because the channel is evolving in response to long term changes in the flow or sediment regime, caused by a change in one of the external basin controls. Examples include incising or aggrading channels and those that are undergoing a change in channel pattern. Alternatively, some bedrock and dryland channels may exist in a permanent state of disequilibrium because it is only during flood flows that adjustments take place.

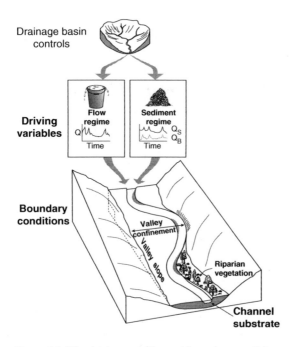

Figure 8.1 The driving variables and boundary conditions that control the form of a channel reach.

In such cases, low flows have little or no influence on the overall channel form. Many empirical relationships have been developed to relate 'regime dimensions' (e.g. channel width or depth), to control variables (e.g. bankfull discharge). It is important to realise that these regime dimensions represent an average and are not applicable to all channel types or flow regimes.

As you saw in Chapter 7, the available stream power along a given reach is determined by the discharge and valley slope. At the sub-reach scale there are spatial variations in energy expenditure, which result from variations in channel shape and resistance to flow. These in turn influence patterns of erosion and deposition. For example, energy and erosion potential are concentrated where the channel narrows. Conversely, flow resistance is increased by obstructions to flow such as boulders, bedforms, bars or woody debris, which can lead to localised deposition. There is therefore two-way feedback between channel form and flow hydraulics – form influences flow and flow influences form.

This point is well illustrated by the work of Ashworth and Ferguson (1986) on a glacially fed braided river in Arctic Norway. An intensive monitoring programme was carried out to make detailed measurements of channel morphology, velocity and shear stress, bedload size and transport rate, and the size of bed material. Ashworth and Ferguson's diagram of the feedbacks between channel processes, morphological changes and sediments is shown in Figure 8.2. Starting at the top left of this diagram is the discharge of the river, which is unsteady (varies over time). The irregular form of the channel creates non-uniform flow conditions over the rough channel bed. As a result, complex spatial variations are seen in velocity, which also changes over time. At any point in the channel, the bed shear stress is determined by the vertical velocity profile (Chapter 6, pp. 87–90). Rates of bedload transport are determined by bed shear stress as well as the size and amount of bed material that is available for transport. As with velocity and shear stress distributions, rates of bedload transport are spatially variable, and also change with time. Bedload transport may maintain the existing channel shape, size and pattern. Alternatively, channel form can be modified as a result of scour, fill and possible lateral migration. The nature of such changes is spatially variable, and in turn feeds back to influence the velocity distribution within the channel. Bedload transport also governs the size distribution and structure of bed sediments through selective entrainment and transport (Chapter 7). The character of the bed material determines the roughness of the channel, in turn affecting the velocity distribution in the channel.

Driving variables

Flow regime

The flow in natural river channels is unsteady, fluctuating through time in response to inputs of precipitation to the drainage basin. Characteristics of the flow regime, discussed in Chapter 3, include seasonal variations, flood frequency–magnitude relationships and the frequency and duration of low flows. Since discharge influences stream power, velocity and bed shear stress, the

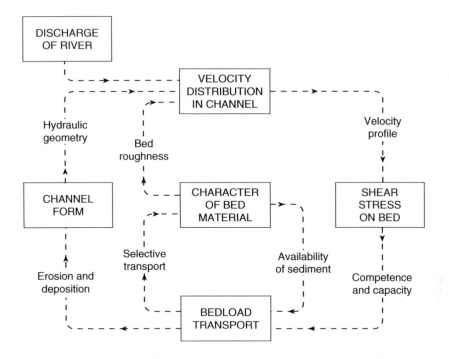

Figure 8.2 Interrelationships amongst form, flow and sediments in active gravel-bed rivers. After Ashworth and Ferguson (1986).

characteristics of the flow regime have an important influence on channel form.

Of morphological significance is the bankfull discharge. This was defined in Chapter 3 and is the discharge at which the channel is completely filled with water. The bankfull discharge marks a morphological discontinuity between within-bank and out-of-bank flows. Since the flow in natural channels is unsteady, the bankfull discharge provides a representative flow. As discussed in Chapter 3, channels are shaped by a range of flows. The geomorphological effectiveness of a given flood depends not only on its size, but also on the frequency with which it occurs. Large floods can carry out a considerable amount of geomorphological work. However, their comparative rarity means that the cumulative effect of smaller, more frequent flows may be more significant in shaping the channel. Box 3.3 provides a fuller explanation of the magnitude and frequency of channel-forming flows.

Bankfull discharge (or the equivalent bankfull width) has often been used in developing statistical relationships between discharge and channel form parameters. It is important to realise that bankfull discharge is actually quite difficult to define and that its frequency of occurrence varies considerably between different rivers (see Chapter 3, p. 32).

Sediment regime

The supply of sediment varies through time. It is not only the volume of sediment that is important but also its size distribution. As you will see later in this chapter, there are significant differences in the behaviour and morphology of bedload, suspended load and mixed load channels.

Fluctuations in the volume and size of sediment are brought about by variations in sediment supply from the drainage basin (Chapter 4) and processes of sediment

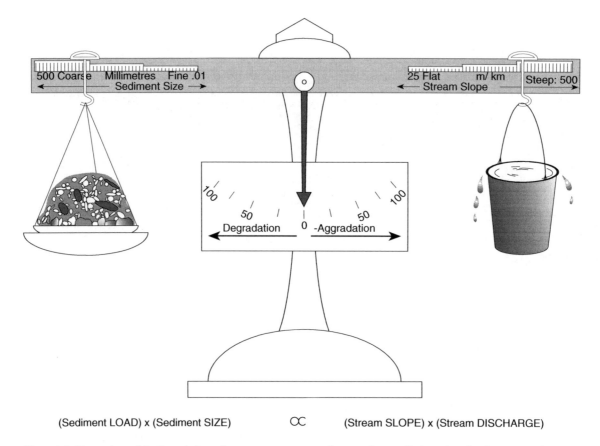

(Sediment LOAD) x (Sediment SIZE)   ∝   (Stream SLOPE) x (Stream DISCHARGE)

Figure 8.3 Illustration of the Lane balance between stream power (stream slope × discharge) and sediment supply (sediment load × sediment supply). This diagram, which originated as an unpublished drawing by W. Borland of the US Bureau of Reclamation, has been adapted from Brierley and Fryirs (2005).

transfer through the channel network (Chapters 5 and 7). As with the flow regime, it is the processes in the drainage basin, upstream from a given reach, that influence sediment supply.

The balance between stream power and sediment supply

There is an important balance between the supply of bedload at the upstream end of a channel reach and the stream power available to transport it. This is known as the Lane balance, having first been described as a qualitative equation by Lane in 1955. This balance is illustrated schematically by the pair of scales shown in Figure 8.3.

The left hand side of the scales represents the volume and size of sediment supplied to a channel reach over a given period of time. Balanced against this is the stream power available to transport it. This is determined both by the volume of water that enters the reach (over the same time period), and by the slope over which it flows.

If the stream power is exactly sufficient to transport the sediment load, both sides of the scales are in balance and there is no net erosion or deposition along the

Plate 8.1 A degrading reach along the Connecticut River, United States, where severe bank erosion is threatening a cornfield. Photograph copyright © Tim McCabe, NRCS.

reach. This is not to say that there is no erosion or deposition whatsoever, because these processes do occur at a localised scale in response to local variations in hydraulic conditions. Rather it means that, on balance, neither erosion nor deposition will predominate.

An imbalance will occur if there is an increase in the volume or calibre of the sediment load in relation to the available stream power (sediment calibre is important because it determines the flow competence required to transport it). This means that there is insufficient stream power to transport all the sediment, with the result that the excess is deposited along the reach. In this case, the balance tips towards **aggradation**, with net deposition occurring along the reach. Aggradation can be triggered in several ways, for example where the sediment supply is increased by upstream channel erosion, mass movement, or human activities such as mining. A particularly dramatic example of mining-induced aggradation along the Ok Tedi River in Papua New Guinea is shown in Colour Plate 16. Aggrading channels are characterised by numerous channel bars in a wide, shallow channel. Deposition within the channel may lead to the channel bed becoming elevated above the surface of the flood-plain. This, together with reduced channel capacity, increases the incidence of flooding and also promotes channel migration.

A different situation arises when the stream power exceeds what is needed to transport the sediment load through the reach. This excess energy has to be expended somehow, so it is used to entrain sediment from the bed and erode the channel boundary. In this case **degradation** predominates. An example of a degrading channel is shown in Plate 8.1. Degradation can be caused by an increase in discharge, perhaps caused by an increase in flood frequency, or by a decrease in sediment supply. This can occur downstream from dams or where gravel mining has removed sediment from the river bed (Chapter 5).

The Lane balance is simplistic because much depends on the calibre of bed sediment within the reach. For example, no degradation can occur in a boulder-bed stream if the bed sediment is too coarse to be moved by the available stream power. This can be true even if the stream power exceeds the sediment supplied at the upstream end of the reach. Even when degradation does occur, another limitation of the equation is that it does not tell us *where* within the reach erosion will occur (Simon and Castro, 2003). This means that the equation cannot be used to predict the actual nature of channel change. For example, if the channel bed is more resistant to erosion than the banks, bank erosion is likely to be an initial adjustment. However, in a sand-bed channel with cohesive banks it is more likely that an initial adjustment would be scouring of the bed (Simon and Castro, 2003). Resistance to erosion can be highly variable within a given reach, as can the specific stream power along that reach. This gives rise to spatially complex adjustments along the reach, even if there is net aggradation or net degradation along the reach as a whole.

Boundary conditions

Valley slope

This refers to the downstream slope of the valley floor (as opposed to the slope of the channel itself) and determines the overall rate at which potential energy is expended along a given reach. The valley slope imposed on a given reach of channel is determined by a combination of factors including tectonics, geology, the location of the reach within the drainage basin and the

Plate 8.2 A confined valley setting, Hundar Gorge, Ladakh, northern India. The figure to right of centre provides an idea of scale.

long-term history of erosion and sedimentation along the valley. Although the overall energy available along a given reach is largely determined by the valley slope, it is possible for adjustments to occur that increase flow resistance at different scales (channel resistance, form resistance and boundary resistance). Different types of channel and floodplain morphology are associated with low, medium and high-energy environments.

Valley confinement

A channel may be defined as confined, partly confined, or unconfined, depending on how close the valley sides are. The degree of valley confinement is important for several reasons. In **confined** settings (Plate 8.2) channel adjustments are restricted by the valley walls, which also increase flow resistance. In addition, valley width influences the degree of slope–channel coupling that exists. Inputs of sediment from mass movements and other slope processes may exceed transport capacity, in turn influencing channel form. The episodic nature of mass movements means that these contributions can vary considerably over time.

In **partly confined** settings (an example is shown in Colour Plate 10), some degree of lateral migration and floodplain development is possible. However, where the river comes against the valley wall or hillslope it is prevented from migrating further, which can lead to the development of over-deepened sections of channel. Stream power is also concentrated within the narrow valley and sections of the floodplain surface may be stripped during major floods.

Where the hillslopes are a long way from the channel and have relatively little influence in contributing to the channel load, the channel is described as **unconfined** (Colour Plate 4). Typically these settings are found in the lower reaches of rivers where there is very little interaction between channel and hillslopes.

Channel substrate

Considerable variations are seen in the form and behaviour of channels developed in different substrates. The substrate determines how resistant the channel is to the erosive force of the flow. It also influences boundary roughness, and therefore flow resistance (Chapter 6, pp. 79–80).

Alluvial channels formed in sand and gravel are generally more easily adjusted than those with cohesive silt and clay substrates. This is because the individual particles can be entrained at relatively low velocities, so non-cohesive substrates tend to be associated with wider, shallower cross-sections and faster rates of channel migration. Bedrock and mixed bedrock-alluvial channels are influenced over a range of scales by various geological controls.

Riparian vegetation

Vegetation on the banks and bed of river channels controls channel form in various ways. It often acts to protect and strengthen the banks, and research has shown

Plate 8.3 The Slesse Creek, British Columbia, in 1940, prior to logging operations. BC air photo BC207-55, copyright © Province of British Columbia. All rights reserved. Reprinted with permission of the Province of British Columbia.

as well as by woody debris (fallen trees and branches) that enters the channel from the banks.

An interesting example of the influence of riparian vegetation on channel form is provided by the Slesse Creek, British Columbia, Canada, and is reported by Millar (2000) and MacVicar (1999). The Slesse Creek drains an area of 170 km^2 within the Fraser River basin, flowing southwards from the United States into British Columbia. The reach shown in Plate 8.3 is approximately 2 km north of the US–Canadian border. This photograph was taken in 1940 and shows a stable meandering channel with a width of approximately 30 m and bordered by native forest vegetation. During the 1950s and 1960s logging of this stream bank vegetation took place along the British Columbia part of the channel. The second photograph (Plate 8.4) shows the same

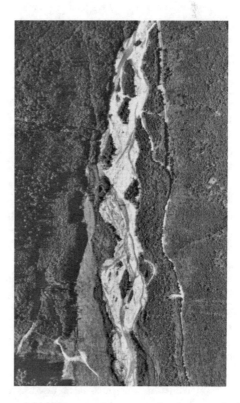

Plate 8.4 The Slesse Creek in 1993, after the removal of riparian forest. BC air photo BC93026-77, copyright © Province of British Columbia. All rights reserved. Reprinted with permission of the Province of British Columbia.

that a dense network of roots can increase erosion resistance by more than a factor of ten. As a result, channels with vegetated banks are often narrower than those with non-vegetated banks under similar formative flows. This effect is most marked for densely vegetated banks (Hey and Thorne, 1986). Flow resistance can also be increased by vegetation growing on the bed and banks, reach in 1993. Subsequent to the removal of forest, the channel has developed an unstable braided form, widening to approximately 150 km. Millar (2000) suggests that these changes are mainly attributable to reduced bank stability, resulting from logging along river banks. This is because there has been no logging in the upstream (US) part of the drainage basin, which is a protected conservation area, so it is assumed that there has been little change in the flow or sediment regimes.

Downstream changes

Downstream changes in slope, discharge, valley confinement, sediment supply and particle size give rise to different balances between erosion and deposition along different parts of the profile. This leads to downstream changes in channel and floodplain morphology. In general terms, the cumulative supply of sediment increases downstream but the available energy decreases.

Figure 8.4(a) shows how selected channel parameters change through the sediment production, transfer and deposition zones. The discharge in most river channels increases in a downstream direction, as a progressively larger area is drained. In order to accommodate the growing volume of flow, channel dimensions (width and depth) typically increase downstream, and are often accompanied by a slight rise in velocity. The way in which these parameters change with increasing discharge can be described by the hydraulic geometry of the channel. Further information on hydraulic geometry relationships is provided in Box 8.1.

Downstream reductions in bed material size reflect differences in the way in which coarse and fine sediment are transferred along the channel. In contrast to the relatively localised transport of bedload particles, fine material, carried in suspension, is transported over much greater distances (Chapter 5, p. 60).

Observations show that there is a general decline in sediment size along the channel. The main causes of this downstream reduction are widely recognised as being abrasion and selective transport (Rice and Church, 1998). Abrasion refers to the reduction in size of individual particles by chipping, grinding and splitting.

Physical and chemical weathering processes are also significant in the pre-weakening of individual particles. Selective transport refers to the longer travel distances associated with smaller grains, which are more mobile. The rate of reduction in sediment size varies considerably and downstream *increases* are often observed at several locations. The downstream decrease in sediment size is often disrupted by inputs of coarser material. These include material from bank erosion, inputs from tributaries, and colluvial material. Material entering the main stream from tributaries is typically coarser than that in the main channel (Knighton, 1998). This causes a sudden increase in sediment size followed by a progressive fining further downstream. Complex patterns of downstream size reduction are seen where slope-channel coupling is strong and non-alluvial supplies are dominant. These include contributions from hillslopes (e.g. mass movements), the erosion of bedrock outcrops and glacial material (Rice and Church, 1998).

The channel slope (represented by the thick line in Figure 8.4a) is typically steepest in the headwaters, becoming gentler in the lower reaches. The resulting long profile of many rivers is concave in shape, although the degree of concavity varies. Downstream increases in discharge, together with a decrease in bed material size, mean that the load can be transported over progressively shallow slopes. Exceptions to this are seen in arid and semi-arid regions, where downstream conveyance losses and high rates of evaporation lead to a downstream reduction in discharge. In this case a straight or convex profile may develop, since increasingly steep slopes are needed to compensate for the downstream reduction in flow.

Irregularities are often seen in the long profile, for example flatter sections are associated with lakes and reservoirs, and steeper sections at the site of waterfalls (see Figure 8.4b). In addition, there is often a change in the channel slope where tributaries join the main channel, because of the sudden increase in discharge. In tectonically active areas, where rates of uplift may be similar to erosion rates, rivers are in a state of dynamic equilibrium constantly trying to 'catch up' with tectonically driven changes. It takes time for a concave profile to develop, so the *overall* shape of long

Box 8.1

HYDRAULIC GEOMETRY RELATIONSHIPS

Discharge varies over space, usually increasing in a downstream direction, and over time in response to inputs of precipitation. Hydraulic geometry describes the way in which flow velocity, width and depth change in response to these discharge variations.

Downstream hydraulic geometry refers to the changes in width, depth and flow velocity that accompany downstream increases in discharge. An example is provided in Figure 1 for three locations along a channel (top left of figure). Section A is located in the headwaters, B in the middle reaches and C further downstream. The channel cross-section at each location is shown on the right-hand side of the figure. Looking at these, the most obvious thing is the downstream increase in channel size. Notice also how the width of the channel increases faster than the depth. These downstream changes are represented by the three graphs on the right-hand side of the figure, where width, depth and velocity are each plotted against mean annual discharge.

The slope of each line indicates the rate at which that variable changes downstream. As mentioned before, width increases faster than depth. The slope of the velocity graph is very gentle because downstream increases in velocity are relatively small. In some cases, velocity shows little downstream change, or even decreases slightly. The relationships shown by the graphs can be described mathematically by the equations:

$$w = aQ^b$$

$$d = cQ^f$$

$$v = kQ^m$$

where w, d, v and Q are width, depth, velocity and discharge and a, c, k, b, f and m are numerical constants.

These equations are examples of a type of relationship called a power function, with the values for the numerical constants describing the relationship for each. The relationships are slightly different for each river, which will have its own unique values for a and b,

defining the downstream relationship between width and discharge; c and f, for depth and k and m for velocity. The exponents (powers) b, f and m define the slope of each graph and, since discharge is equal to the product of width, depth and velocity, from basic algebra it follows that $b + f + m = 1$. Power functions appear as a straight line when plotted using a logarithmic scale. Average values for b, f and m are generally reported to be 0.5, 0.4 and 0.1 respectively.

From casual observations of your local river you have probably noticed how the depth, width and velocity of flow all increase as the river rises. **At-a-station hydraulic geometry** relationships describe variations at one particular cross-section as discharge changes through time. This is illustrated in Figure 1 for section B, with B1, B2 and B3 showing low, moderate and mean annual flood discharges. The corresponding schematic graphs are on the left hand side of the figure, using average values for the exponents ($b = 0.26$, $f = 0.40$ and $m = 0.34$).

There is actually a fair amount of variation between different cross-sections. In wide, shallow channels the width tends to increase faster than the depth (b is greater than f), while in deeper, narrower channels it is the depth that increases at a faster rate (b is less than f). Changes in velocity are greatly influenced by elements of flow resistance at a given cross-section. In general where flow resistance is high, as in a braided channel where bars increase roughness, velocity tends to increase at a slower rate (lower value of m) than in channels with less flow resistance.

Downstream hydraulic geometry relationships may be applied to identify bankfull stage and channel dimensions in ungauged basins. This assists in the design of stable natural channels for river management (Rosgen, 1994). These relationships can also be useful in studies of the effects of land-use changes on channel dimensions (Gordon *et al.*, 2004). However, it is important to realise that there are a number of limitations associated with hydraulic geometry relationships.

Box 8.1

HYDRAULIC GEOMETRY RELATIONSHIPS—CONT'D

In particular, care must be exercised when applying empirically-derived relationships beyond the range of data for which they were derived. The underlying data are representative only of the channel type and environment for which the measurements are made. This means that the relationships should not be applied beyond this, for example in predicting stable channel dimensions for a different type of river environment.

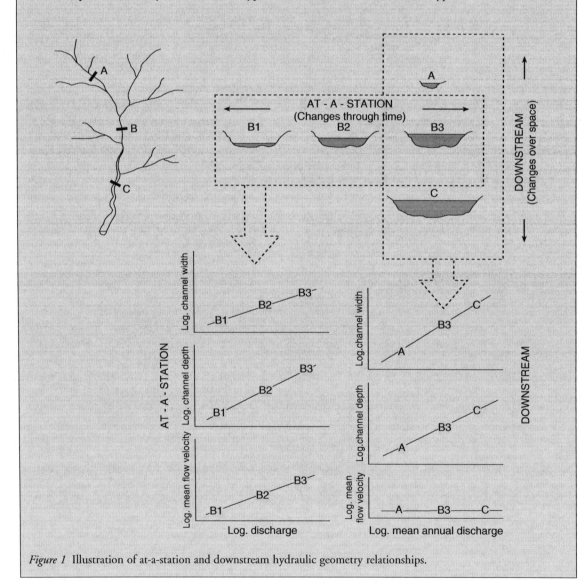

Figure 1 Illustration of at-a-station and downstream hydraulic geometry relationships.

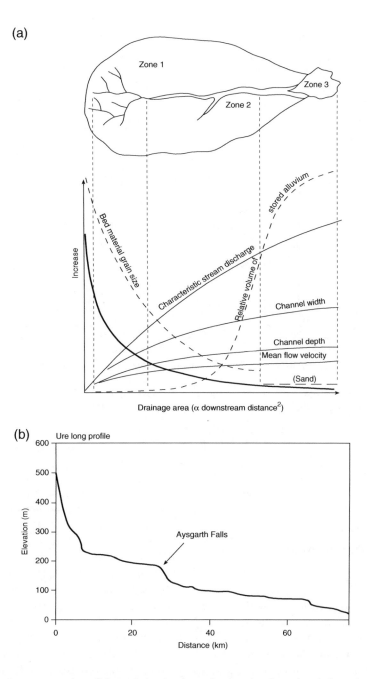

Figure 8.4 (a) Schematic representation of the variation in channel properties through a drainage basin. After Church (1992) and based on a concept of Schumm (1977). (b) Long profile of the River Ure, North Yorkshire, England. The Aysgarth Falls are a closely spaced series of three low waterfalls. After Coulthard *et al.* (2005).

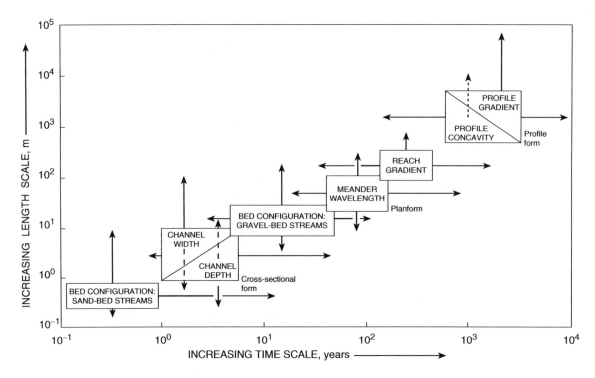

Figure 8.5 Schematic diagram of the time scales of adjustment of various channel form components with given length dimensions in a hypothetical basin of intermediate size. After Knighton (1998).

profiles in tectonically active regions tends to be straight rather than convex.

CHANNEL ADJUSTMENT

Time scales of adjustment

Different components of a channel's morphology (e.g. bedforms, cross-sectional shape, slope) change over different time scales (Figure 8.5). This is because some components are more readily adjusted than others. For example, bedforms in a sand-bed channel are rapidly modified by a wide range of flows. Adjustments to channel width and depth take place over months to years, planform adjustments occur over tens to hundreds of years, while changes in the long profile may take thousands of years. Morphological adjustments therefore tend to lag behind the changes that cause them. This means that it can be difficult to link processes of flow and sediment transport with channel dimensions and form.

Channel form is directly controlled by flow regime and sediment supply. This chapter will deal with relatively short-term adjustments that are directly influenced by the flow and sediment regimes. Chapter 9 will examine longer-term changes that are brought about by changes in the external basin controls (climate, tectonics, base level and human activity). These all act as controls on the flow and sediment regimes and, through a complex sequence of adjustments, lead to long-term changes in channel form and behaviour.

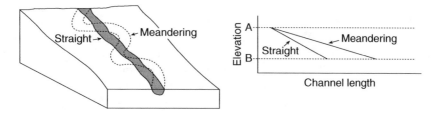

Figure 8.6 The development of meanders between two points, A and B, increases channel length. This reduces the channel slope between A and B.

How adjustments are made

Channel form and behaviour reflect the driving variables and boundary conditions influencing a given channel reach. These controls also influence the ways in which channel adjustments are made. There are potentially four **degrees of freedom**, or variables, that can be modified: channel cross-section, slope, planform and bed roughness.

Modifications to the cross-sectional size and shape are associated with changes in width and depth of the channel by processes such as bank erosion, incision of the bed, or aggradation. Channel slope can be adjusted in different ways. Negative feedback reduces the slope of steeper sections by erosion, and the slope of flatter sections is increased by deposition. Increases or decreases in channel length also affect channel slope, as illustrated in Figure 8.6. There are several different types of channel planform adjustments. These include lateral migration, meander bend development, reworking of bars, and even wholesale shifts of the channel to a new course. Finally, changes in bed roughness are brought about when the channel rearranges bed material, for example, in sand-bed channels, where bedforms are modified in response to changes in flow conditions. Mutual interrelationships exist between these variables, with adjustments made to one affecting one or more of the others. For instance, the formation of a meander cut-off alters the channel planform as well as increasing channel slope.

The influence of the driving variables and boundary conditions often reduces the degrees of freedom that a particular channel has. In the case of the mixed bedrock-alluvial channel shown in Colour Plate 3, depth increases are greatly restricted by the rock bed of the channel. On the other hand, the alluvial banks allow the channel to be widened much more easily. However, reductions in cross-sectional size by deposition may be limited if the channel has degradational tendencies.

CHANNEL GEOMORPHIC UNITS

Geomorphic units are features that form at the sub-channel scale and can be erosional or depositional in origin. Distinctive assemblages or groupings of geomorphic units characterise the different channel types introduced in Chapter 1. For instance, braided channels contain numerous mid-channel bars, while bedrock channels are associated mainly with erosional features such as potholes and bedrock steps, although bars can also form if sufficient bed sediment is available. Geomorphic units also affect hydraulic processes, and provide a range of different habitats for in-stream flora and fauna.

Bars

Bars are in-channel accumulations of sediment which may be formed from boulders, gravel, sand or silt. Bars can be divided into two broad groups: **unit bars** and **compound bars** (Smith, 1974). Unit bars are relatively simple bar forms whose morphology is mainly determined by processes of deposition (Ashmore, 1991). The evolution of these simple bar forms into more complex forms is described by Smith (1974), who made observations of the Kicking Horse River, British Columbia, Canada.

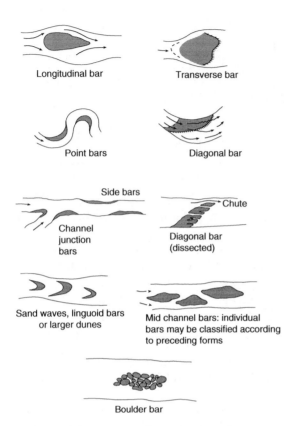

Figure 8.7 Types of channel bar. Adapted from Morisawa (1985), Church and Jones (1982) and Church (1992).

found where there is an abrupt channel expansion, and downstream from confluences (Church and Jones, 1982). Transverse unit bars are not usually attached to the banks (Robert, 2003). The channel junction bars shown in Figure 8.7 are transverse bars that are associated with the flow separation that occurs at channel confluences.

Point bars are a feature of most meandering channels and form on the inside of meander bends as a result of the secondary flow patterns that are associated with flow in curved channels. Point bars are elongated in the direction of flow, with a steep outer face.

Diagonal bars are common in gravel-bed channels (Robert, 2003). These are bank-attached features that run obliquely across the channel. Diagonal bars may have a steep downstream front.

Both longitudinal and transverse bars are closely related to mid-channel bars. The compound mid-channel bars that characterise braided channels often have a complex history (see Colour Plates 6 and 7 for examples of these compound bars). Two terms that are commonly used to describe complex bar forms are medial (or lingoid) bars and lateral bars (Robert, 2003). **Medial bars** are symmetrical, detached from the banks and have a characteristic lobate shape. **Lateral bars** are attached to one bank and have an asymmetric shape. Both types of compound bars have complex evolutionary histories.

Boulder bars form in channels that are dominated by coarse bedload. As you will see later in this chapter, different morphologies are associated with the islands that are associated with anabranching channels. These include sand ridges, excavated islands, bedrock bars and vegetated bars with a bedrock core.

Compound bars have a more complex history, having been shaped by many episodes of erosion and deposition. When erosion occurs, the basic shape of the bar is trimmed and dissected.

Church and Jones (1982) recognise four main types of unit bars. These are illustrated in Figure 8.7. **Longitudinal bars** are elongated in the direction of flow. They form in the centre of the channel, typically where the channel is relatively wide. Bar growth is brought about by the accumulation of finer material, both in an upwards and in a downstream direction (Church and Jones, 1982). Longitudinal bars tend to taper off in a downstream direction (Robert, 2003).

Transverse bars are lobe shaped (lobate) with relatively steep downstream faces. They are commonly

Benches

Benches are flat-topped, elongated, depositional features that form along one or both banks of channels. They are typically found on the inside of bends and along straight reaches, and are intermediate in height between the level of the channel bed and floodplain (Figure 8.8a). In bedrock and boulder-bed channels a boulder berm (bench composed of boulders) may form at the edge of the channel.

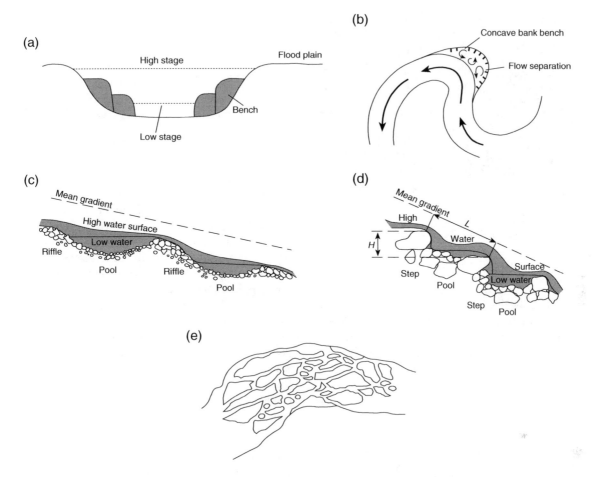

Figure 8.8 (a) Cross-sectional view of channel benches. (b) Concave bank bench. (c) Riffle–pool sequence. After Knighton (1998). (d) Step–pool sequence. After Knighton (1998). (e) Bedrock bars.

Benches can also form where flow separation occurs at the outer (concave) bank of tightly curving meander bends. This results in deposition and is illustrated in Figure 8.8(b).

Erskine and Livingstone (1999) have observed sequences of adjacent benches along a bedrock-confined channel in the Hunter Valley, New South Wales, Australia. Rivers in this region have a very high flow variability, and each bench is associated with a different flow frequency. These benches are often eroded by catastrophic floods but are subsequently reconstructed by lower magnitude floods.

Riffle–pool sequences

The terms riffle and pool come from trout angling and refer to large-scale undulations in the bed topography. They are commonly found in gravel-bed channels with low to moderate channel slopes but do not tend to form in sand- or silt-bed channels (Knighton, 1998). The difference between riffles and pools is most obvious at low stages, when the flow moves rapidly over coarse sediment in the relatively steep riffle sections and more slowly through the deeper pools (Plate 8.5). The spacing from pool to pool, or riffle to riffle, is related to the width of the channel (and hence flow discharge).

Plate 8.5 Riffle and pool on the upper River Ure, North Yorkshire, England. The riffle is marked by the broken water surface.

In most cases this is between five and seven times the channel width (Keller and Melhorn, 1978). A longitudinal section through a riffle–pool sequence is shown in Figure 8.8(c). This illustrates the differences in bed slope, bed material size and the slope of the water surface at high and low flows. At higher flows, the differences between riffles and pools are less obvious, with less variation in the water surface slope.

Riffle–pool sequences are found in straight, meandering and braided reaches. Analogous features are sometimes seen in ephemeral channels as regularly spaced accumulations of relatively coarse sediment, although there is little variation in the bed topography (Leopold *et al.*, 1966). In ecological terms, both riffles and pools provide important habitats. For example, certain species of fish lay their eggs in the spaces between the coarse gravels in riffles, while pools provide shelter and a suitable habitat for rearing young.

Various theories have been put forward to explain how riffle–pool sequences are maintained. Keller (1972) introduced a theory of velocity reversal. This suggests that the flow velocity increases at a faster rate in pool sections than in riffles as the discharge approaches bankfull. The higher shear stresses that develop in the pools lead to scouring of coarse material, which is deposited immediately downstream to form riffles. However, there is conflicting evidence to support this theory.

Several researchers have shown that pools have a larger cross-sectional area of flow than riffles during most flow conditions. In order to ensure continuity of flow, pools should therefore have lower cross-sectional velocities (see Chapter 6, pp. 76–77). For example, Carling (1991) made observations on the River Severn, England. These indicated that neither the cross-sectional average velocity nor the near-bed shear velocity were noticeably greater in pools than riffles during overbank/near-overbank conditions. Instead, there was a tendency for average hydraulic variables in riffles and pools to become more similar as the discharge increased.

Other theories have also been put forward. For example, field and laboratory measurements have shown that riffle surfaces tend to experience more turbulent flows. As a result, a tightly packed and interlocked bed surface develops at riffles. This is brought about by the vibration of particles and occasional particle transport during relatively low flows. In contrast, pools experience less near-bed turbulence during low flows and do not develop the same type of resistant bed structure (Robert, 2003). This means that critical bed shear stresses for sediment entrainment are higher in riffles than in pools. The riffles therefore tend to be maintained as topographic high points, while scouring occurs at pools (Robert, 2003).

Steps and pools

Steps and pools (Figure 8.8d and Plate 8.6) often characterise steep, upland channels and have been observed in a wide range of humid and arid environments. The steps are formed from coarser material and form vertical drops over which the flow plunges into the deeper, comparatively still water of the pool immediately downstream. Steps are relatively permanent features and consist of a framework of larger particles that is tightly packed with finer material. In forested catchments, woody debris has been observed to form part of the structure of steps. Steps and pools can also form in bedrock channels.

Like riffles and pools, step–pool sequences are most apparent during low-flow conditions as they tend to be drowned out at higher flows. It is also during low-flow conditions that step–pool systems offer the most flow

Plate 8.6 Steps and pools, Julian Alps, Slovenia.

resistance. There is a considerable dissipation of energy as flow cascades over each step and enters the relatively still pools (Bathurst, 1993).

The spacing of steps and pools has been widely reported as being, on average, two to three times the channel width. Pools also tend to become more closely spaced as the slope increases. The height of steps appears to increase with the size of the bedload (Chin, 1999).

Channels in which step–pool sequences form typically have a wide range of sediment sizes, from fine gravel to large boulders. Laboratory-based simulations indicate that step–pool sequences probably form during large floods, which mobilise the coarsest sediment. One theory suggests that, when the coarsest 'keystones' come to rest, they act as a barrier, leading to the accumulation of finer sediment. Downstream from this, the flow of water over the step scours a pool (Knighton, 1998).

Rapids and cascades

Like step–pool sequences, these are associated with steep channel gradients. Rapids are characterised by transverse, rib-like arrangements of coarse particles that stretch across the channel, while cascades have a more disorganised, 'random' structure. Rapids and cascades are stable during most flows because only the highest flows are competent to move the coarser cobbles and boulders that form the main structure.

Potholes

These deep, circular scour features are formed in bedrock channel reaches by abrasion (Plate 7.1). Processes of formation are described in Chapter 7 (p. 96).

Bedrock bars

In incised bedrock channels, the flow sometimes moves around bedrock bars (Figure 8.8e). These form when multiple sub-channels are incised into the bedrock substrate, leaving 'islands' or bedrock bars between them. Bedrock bars may form the core of a bedrock-alluvial bar, which becomes covered by a layer of sediment on which vegetation becomes established.

FLOODPLAIN MORPHOLOGY

Processes of floodplain formation

The morphology of floodplains is intimately linked with the form and behaviour of the river channels that shape them. Various processes of deposition, reworking and erosion are involved in the formation and development of floodplains. Sediment accumulates on floodplain surfaces by various processes of accretion, the main ones being vertical, lateral and braid bar accretion (Nanson and Croke, 1992). **Lateral accretion** deposits are laid down by migrating rivers, which erode into the floodplain and lay down sediment in their wake. The accretion of point bar deposits can sometimes be seen as a series of concentric ridges on the inside of bends called **meander scrolls**. **Braid bar accretion** occurs when bars are abandoned and gradually become incorporated into

Plate 8.7 High-energy non-cohesive floodplain, Tien Shan Mountains, Kyrgyzstan. The channel is approximately 3 m wide. Note the confined setting, steep gradient and coarse sediment. High discharges are associated with the spring snowmelt.

the floodplain deposits. There are various ways in which this can happen, for example when a large flood lays down extensive bar deposits. Alternatively, bars may become abandoned when the main braid channels shift to another part of the valley. **Vertical accretion** deposits are composed of fine material that settles out of suspension when overbank flows inundate the floodplain. The increased area of contact, coupled with the roughness of the floodplain surface, greatly reduces flow velocities, and a thin layer of sediment is draped across the floodplain. This displays a fining-upwards sequence, where the coarser particles, which settle out first, are overlain by progressively finer material. There is also a fining of sediment away from the channel, since only the very smallest particles are carried to the edge of the inundated area. Over a number of years the cumulative effect of overbank flows leads to the development of a vertical sequence of thin layers. Other, more localised, types of accretion can also be identified. For example, **counterpoint accretion** is associated with the deposition of concave bank benches at confined meander bends (see section on channel geomorphic units above). As an over-tightened meander bend migrates, bench deposits become incorporated into the floodplain.

Erosional processes include **floodplain stripping**, where entire sections of the floodplain surface are removed by high-magnitude flood events. Floodplain stripping is most likely to occur in relatively confined valley settings, where floodplain flows are concentrated between the valley walls. Other erosional processes include the formation of flood channels, which carry water during overbank flows. **Avulsion** involves a shift in the position of a channel and is a common process in braided reaches where the flow frequently abandons and reoccupies sub-channels. Avulsion can also involve the diversion of flow into a newly eroded channel cut into the floodplain. This type of avulsion is important in the development of anabranching channels.

The morphology and development of floodplains is controlled by the driving variables and boundary conditions. An important balance exists between the shear stress exerted by the flow and the resistance of the floodplain to erosion. Shear stress is closely related to specific stream power (Chapter 7, pp. 93–94), and therefore to such controls as flow regime, valley slope and valley confinement. On the other side of the balance, resistance to erosion is largely determined by the cohesiveness of the floodplain sediments. An energy-based floodplain classification was proposed by Nanson and Croke (1992). This recognises three main classes of floodplain:

- *High-energy non-cohesive floodplains* are typically found in steep upland areas where the specific stream power in the channel at bankfull flow exceeds 300 W m^{-2}. An example is shown in Plate 8.7. Lateral migration is often prevented by the coarseness of the floodplain sediment, which builds up vertically over time. These floodplains are disequilibrium features that are partly or completely eroded by infrequent extreme events.
- *Medium-energy non-cohesive floodplains* are formed from deposits ranging from gravels to fine sands. Specific stream power ranges from 10 to 300 W m^{-2}. The main processes of floodplain construction are lateral point bar accretion (meandering channels) and braid bar accretion (braided channels). These floodplains are typically in dynamic equilibrium with the annual to decadal flow regime. Some of the features associated with

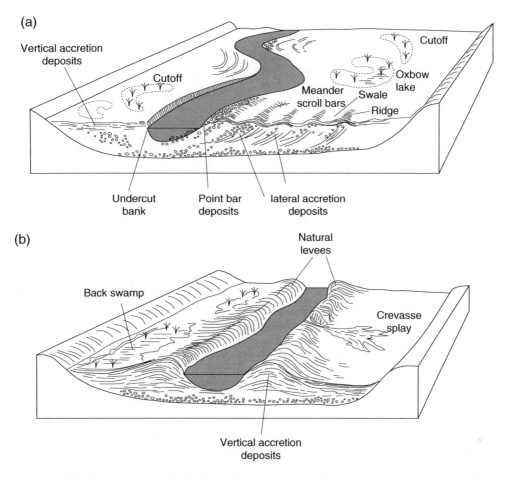

Figure 8.9 Some of the features associated with (a) medium-energy non-cohesive meandering floodplains, (b) low-energy cohesive floodplains. Adapted from Nanson and Croke (1992).

medium-energy non-cohesive floodplains are illustrated in Figure 8.9(a).

- ***Low-energy cohesive floodplains*** are usually associated with laterally stable single thread or anastomosing channels. Formed from silt and clay, the dominant processes are vertical accretion of fine-grained sediment and infrequent channel avulsions. Specific stream power at bankfull stage is generally less than 10 W m^{-2}. Features associated with low energy cohesive floodplains are shown in Figure 8.9(b).

These classes can be further subdivided, mainly on the basis of floodplain forming processes. They can also be related to the downstream reductions in stream power and sediment size discussed on pp. 124–128.

Floodplain geomorphic units

Levees

Levees are elongated, raised ridges that form at the channel–floodplain boundary during overbank flow events (Figure 8.9b). Moving across the boundary from channel to floodplain, there is a sudden loss of momentum because of the interaction between fast channel flow and slow floodplain flow (see Chapter 6, pp. 88–90 and Figure 6.10). This results in the preferential deposition of material at the edges of the channel. Colour Plate 9 shows an anastomosing river in flood. Levees are clearly visible as the raised strips of land running along the channel margins. The height of levees is scaled to the size of the channel and their presence implies a relatively stable channel location (Brierley and Fryirs, 2005). These natural levees should not be confused with the artificial levees that are constructed along river banks for purposes of flood control (discussed in Chapter 10).

Crevasse splays

Levees can be breached by floodwaters. This may lead to the formation of a crevasse splay, a fan-shaped lobe of sediment deposited when sediment-charged water escapes and flows down the levee (Figure 8.9a). If flow is sufficiently concentrated, a new channel may be cut and deepened by scour.

Backswamps

The build-up of sediment in the channel may mean that the channel is at a higher elevation than the surrounding floodplain. When levees are overtopped, water can enter the lower-lying area on the other side of the levee. This may be a depression or a swamp area characterised by wetland vegetation (Figure 8.9b). Colour Plate 9 shows some good examples of backswamps. These are not exclusively associated with anabranching rivers and can also form at the valley margins of other channel types.

Flood channels

Flood channels are relatively straight channels that bypass the main channel. They have a lesser depth than the main channel and are dry for much of the time, only becoming filled with water as the flow approaches bankfull.

Floodouts

Floodouts are associated with dryland channels. They occur where floodwaters leave the main channel and branch out onto the floodplain in a number of distributory channels. This happens where low gradients, downstream transmission losses and high rates of evaporation lead to a downstream reduction in channel capacity. Channels may re-form downstream from the floodout if flow concentration is sufficient, forming a discontinuous channel. Alternatively the floodout may mark the channel terminus. Floodouts can also form where the channel is blocked by bedrock outcrops, fluvial, or aeolian deposits such as sand dunes (Tooth, 1999).

Meander scroll bars

In some cases, former point bar deposits can be seen in the surface topography of the floodplain as **scroll bars**, with each scroll representing a former location of the point bar (Figure 8.9a). The undulating **ridge and swale topography** that results consists of higher ridges separated by topographic lows called swales. Meander scroll bars can be seen as a series of vegetated ridges on the point-bar deposits in the foreground of Colour Plate 10. Migrating meanders do not always form scroll bars and the surface topography of these deposits may be relatively featureless.

Cut-offs

These are abandoned meander bends that have been short-circuited by the flow. Several examples can be seen in Colour Plate 11. Cut-offs become infilled over time by a process of abandoned channel accretion.

Palaeochannels

Palaeochannels are longer sections of abandoned channel (Colour Plate 18). Like active channels, palaeochannels exhibit a wide range of different planforms. As time goes by, they gradually become infilled by abandoned channel accretion, the degree of infilling reflecting the age of the channel. The rate at which infilling occurs is dependent on factors such as the geometry of the palaeochannel and its position on the floodplain in relation to overbank events.

ALLUVIAL CHANNEL FORM

The previous sections have considered the controlling influence of the driving variables and boundary conditions on channel form and behaviour. Each of these controls varies across a continuous range. For instance, slopes range from steep to gentle, valleys from confined to unconfined, and sediment loads from suspended to mixed to bedload dominated. Many different combinations are possible, leading to the immense variety of fluvial forms and behaviour that is seen globally.

The continuum of alluvial channel types is illustrated in Figure 8.10. In general terms, different channel types exist along an energy gradient, ranging from high-energy braided channels through meandering and straight to low-energy anastomosing channels (a sub-set of anabranching channels). Floodouts and chains-of-ponds are found in low-gradient arid environments, where downstream reductions in discharge result in a dwindling supply of energy. This continuum can be related to the channel controls, since stream power integrates channel slope and flow regime. It also influences the type of load that the channel can carry, which in turn determines the substrate and stability of the channel. Like all channel classifications, the one shown in Figure 8.10 is, by necessity, a simplification of reality. While some channel reaches typify, say, a braided form, many have characteristics that are associated with more than one type. In fact, it has been suggested that channels with an intermediate form might be the norm rather than the exception (Ferguson, 1987).

Although there is a continuum of forms, thresholds do exist between them. For example, there is a meandering–braiding threshold, above which rivers braid and below which they meander. Rivers that are close to this threshold, such as many of those in the South Island of

Figure 8.10 The continuum of variants of channel planform. After Brierley and Fryirs (1992), adapted from Church (1992) and Schumm (1977).

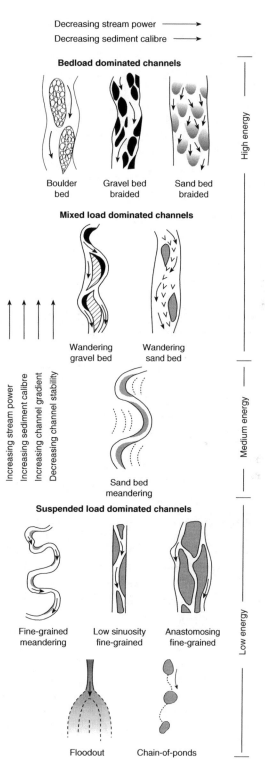

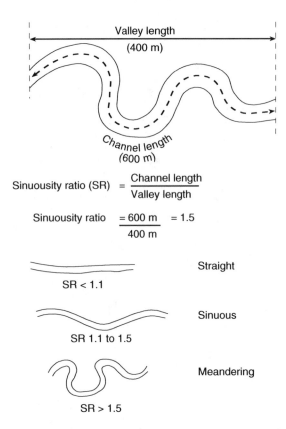

Figure 8.11 Sinuosity ratio definition.

New Zealand, have alternating meandering and braided reaches. For example, the Rangitata River rises in the Southern Alps, passing through a bedrock gorge as it leaves the mountains, before flowing across the Canturbury Plains. Above the gorge, the channel is braided. Below the gorge, and with a reduced sediment load, the Rangitata meanders, in contrast to the other, braided, rivers that flow across the Canterbury Plains. However, further downstream, the river cuts into unconsolidated Pleistocene terrace deposits and reverts to a braided form (Schumm, 1979).

There are a number of management implications associated with such thresholds. Rivers that are close to a threshold may be particularly sensitive to variations in flow or sediment supply, or to channel engineering works, which could result in dramatic changes in form and behaviour. Careful management can (intentionally) convert a transitional braided reach to a more stable meandering form, for example by stabilising banks with vegetation or isolating sediment supplies. In qualitative terms, it is possible to identify rivers that are close to the braided–meandering threshold, since they display a number of forms that can be recognised as transitional. Conversely, a 'typical' meandering channel that does not have transitional features can be assumed to be some distance from the threshold. There have been a number of attempts to define this threshold in quantitative terms, by developing relationships using key variables such as bankfull discharge and channel slope (which control stream power) and the size of bedload. However, because channel form is influenced by so many different variables, this is extremely difficult to do. For a review of quantitative definitions of the braided–meandering threshold see Thorne (1997).

Straight and meandering channels

Most single-channel rivers and streams follow a winding path and straight channels are rare. The **sinuosity ratio** gives an indication of how 'bendy' a channel is and can be worked out by measuring the length of a channel reach and dividing this by the straight line distance along the valley (Figure 8.11). Channels with a sinuosity ratio of less than 1.1 are described as straight, those between 1.1 and 1.5 are sinuous, and meandering channels have a ratio of more than 1.5. Although widely used, these descriptions are somewhat arbitrary, since they are not based on any physical differences.

There is a tendency for the *thalweg*, or line of fastest flow, to shift from side to side along the channel. This is seen even in straight channels, and is often associated with the development of riffles, pools and alternate bars.

Meander geometry

Various methods are used to quantify the geometric characteristics of meandering channels. These are based on measurements that can be made in the field, from maps, aerial photographs and, increasingly, satellite images. The spacing of meander bends, or **meander wavelength (λ)**, can be determined by measuring

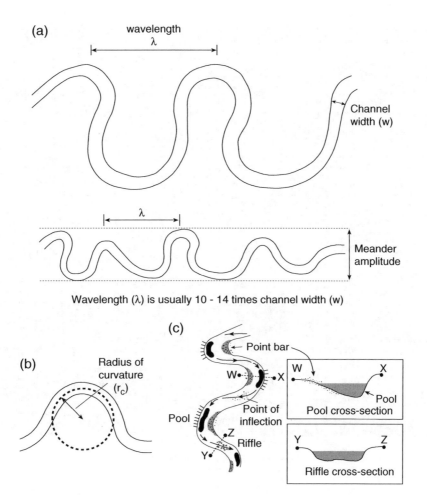

Figure 8.12 Aspects of meander geometry. (a) Meander wavelength. (b) Radius of curvature. (c) Typical channel cross-sections at pools (W to X) and riffles (Y to Z).

the straight-line distance from one bend to the next (Figure 8.12a). Since the distance between successive meander bends varies, a mean wavelength is calculated for several meander bends along the reach of interest. There is a well established relationship between channel width and meander wavelength, which is usually approximately ten to fourteen times the bankfull width. (Chorley *et al.*, 1984). Meander wavelength is more strongly related to channel width than to bankfull discharge. This is because secondary circulation within the

channel, which is significant in meander development, is controlled by channel size (Richards, 1982). Interestingly, a similar relationship is seen for other meandering systems, for example the small supraglacial streams that flow over the surface of glaciers often develop meanders, despite the absence of sediment (Plate 8.8). At a much larger scale, meanders also form in the Gulf Stream of the Atlantic. In both cases, the wavelengths of the meanders are scaled to the width of the flow in the same way as for alluvial channels.

Plate 8.8 A meandering stream incised into the surface of a glacier.

Meander wavelength can also be influenced by the channel substrate, and longer wavelengths are associated with gravel channels than for silt and clay channels of a similar size. The reason for this is that cohesive banks allow the development of a narrower cross-section with tighter bends (Schumm, 1968).

An indication of the 'tightness' of individual bends can be determined by fitting a circle to the centre line of a meander bend (see Figure 8.12b). The radius of this circle is called the **radius of curvature** (r_c). To allow comparison between channels of different sizes, the tightness of bends is usually expressed as the ratio between the radius of curvature and the channel width

at the bend (r_c/w). This ratio is relatively small for tight bends and increases for bends that curve more gradually. Observations have shown that many bends develop an r_c/w ratio of 2 to 3. For bends that are tighter than this, flow separation leads to increased energy losses (Bagnold, 1960).

In cross-section, the form of the channel varies along its length as shown in Figure 8.12(c). An asymmetric cross-section is associated with meander bends since scouring and bank steepening take place at the outside of the bend, while deposition occurs on the inside of the bend. At riffle sections, where the line of fastest flow crosses the channel, the cross-section is more symmetrical.

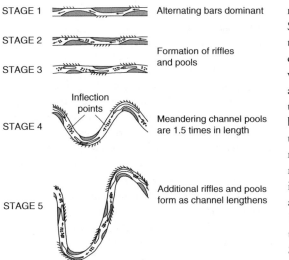

STAGE 1   Alternating bars dominant

STAGE 2

STAGE 3   Formation of riffles
           and pools

Inflection
points

STAGE 4   Meandering channel pools
           are 1.5 times in length

STAGE 5   Additional riffles and pools
           form as channel lengthens

Riffle   Pool   Erosion

Figure 8.13 Transformation of a straight channel with a riffle–pool bed into a meandering channel. After Keller (1972).

The regularity of meander bends varies greatly, with some following rather an irregular path, while others are highly regular. Ferguson (1979) asked whether meanders are regular or random. He concluded that meander form is a compromise between flow behaviour, which tends towards regularity, and the 'random' attributes of floodplains that lead to irregularity. These include variations in floodplain topography and sedimentology.

Why do rivers meander?

It might seem logical that rivers and streams should take the most direct course – a straight line – down the slope of the valley. However, most single-channel rivers show some degree of sinuosity. Over the years, a number of theories of meander formation have been developed on the basis of a growing body of theoretical, field and experimental research. However, no general agreement has been reached.

The influence of channel controls is reflected in the close correlations that exist between meander wavelength,

mean radius of curvature and channel width. Since channel width is related to discharge, the implication is that meanders are scaled to the range of discharges that shape the channel (Box 8.1). Meander wavelength can also be correlated with sediment load and channel slope. Lengthening of the channel reduces the slope (this was illustrated in Figure 8.6). However, bends increase flow resistance because more energy has to be used to move the flow around them. In theory, the most energy-efficient kind of bends are symmetrical meanders. These represent the path of least work, allowing the channel to lengthen, but minimise the associated energy expenditures (Langbein and Leopold, 1966). At the same time, bends allow energy expenditures to be more evenly distributed along the channel. Since flow resistance is greater at riffles than at pools, the additional energy used in turning around bends (where pools are located) balances the increased flow resistance encountered at riffles. However, reach-scale relationships such as these do not actually tell us about the processes involved.

The question of *how* rivers meander is not fully understood but relates to interactions between the flow and the material forming the bed and banks. As water flows through a channel, spiralling secondary flow cells are set up within the flow as a result of boundary resistance (Chapter 6, p. 88). The three-dimensional nature of this flow can be significantly altered by any irregularities in the channel boundary, with the effects of any disturbance being propagated downstream. These flow variations create differences in velocity and shear-stress distributions, and hence patterns of erosion and deposition. This leads to channel-form adjustments, which in turn affect channel flow, resulting in further channel modifications.

Figure 8.13 shows a conceptual model of meander formation that was developed by Keller (1972). During stage 1, alternate bars form on opposite sides of the channel as a result of alternating zones of erosion and deposition. The flow is diverted around these bars, converging as it moves towards the banks and diverging as it moves across the channel. This promotes erosion and pool formation on alternate sides of the channel. Deposition occurs at the crossing points, where riffles form (stages 2 and 3). Erosion continues to be focused

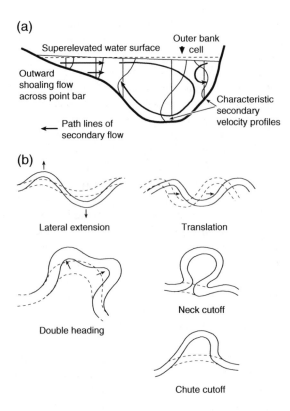

Figure 8.14 (a) Flow in a meander bend. After Markham and Thorne (1992). (b) Styles of meander migration and cut-off formation.

Flow and sediment transport in meander bends

As water flows around a meander bend, the water tends to 'pile up' against the outer bank, resulting in a superelevation of water on that side of the channel (Figure 8.14a). As a result of this localised increase in depth, a pressure gradient develops across the channel (Chapter 6, p. 88). This leads to a compensatory flow of water across the channel bed, from the outside (high pressure) to the inside (low pressure) of the bend. Although relatively weak in comparison with the primary flow in the channel, these secondary flows are significant in moving sediment towards the inside of the bend. Point bar deposits are sorted, with coarser sediment deposited at the base of the bar, while progressively finer sediment is carried up the bar surface. The upper surface of the point bar is often draped with fine vertical accretion deposits laid down during high flows.

A widely recognised feature in meander bends is the zone of high velocity that shifts from the inside to the outside of the bend with increasing distance along the bend. Dietrich *et al.* (1979) made detailed measurements of local boundary shear stresses at various locations along a meander bend in a sand-bed channel. These show how the zone of maximum bed shear stress also shifts from the inside to the outside of the bend. A close correspondence exists between the zone of maximum shear stress and the maximum average flow velocity (Dietrich, 1987). Measurements were also made of bedload transport and rates of bedform migration. These showed that the zones of maximum bedload transport were similar to zones of maximum bed shear stress (Dietrich *et al.*, 1979).

Meander migration

Once meanders have formed, further development often takes place, as individual bends migrate, by erosion of the outer (concave) bank and compensatory deposition on the point bar at the inside of the bend (Plate 8.9). There are various ways in which meanders can migrate, some of which are illustrated in Figure 8.14(b). **Lateral extension** occurs during the formation of meanders and has the effect of lengthening the channel

at the banks, leading to the development of bends (stage 4). As these grow, and the channel extends, new riffles and pools form so that the spacing remains between five and seven times the channel width (stage 5). However, in the case of low-energy channels with resistant banks, there is no progression beyond the straight channel of stage 3. This is because the eroding force of the flow is insufficient to overcome the resistance of the banks. Lateral migration and bend development therefore cannot take place.

The Keller model is not universally applicable because meanders can still form in channels without riffles and pools. While they are commonly observed in gravel bed channels, riffles and pools do not tend to form in sand and silt-bed channels.

Plate 8.9 Meander bend, showing how the fastest flow shifts to the outside of the bend (note cut bank), with deposition on the inside of the bend in the form of a point bar, River Ure, North Yorkshire, England.

and increasing the amplitude of meander bends. The effect of valley confinement on meander development can be seen in Plate 8.10. In this case, the lateral extension of the meander bends is restricted by the valley walls. **Confined meanders** such as these can also form where migration is restricted by rock outcrops or artificial structures such as roads and railway embankments. **Translation** occurs when the fastest flow erodes the bank downstream from the bend apex, resulting in down-valley movement. Meander bends often develop an asymmetric planform when one limb of the meander migrates at a faster rate than the other, a situation that may be caused by variations in bank resistance along the channel.

Meanders can develop more complex forms, for example bends with two apices (double heading) or where lobes form on existing bends (Hooke and Harvey, 1983).

However, meander bends do not continue to grow indefinitely, or the channel slope would become too gentle to allow transport of the sediment load. Instead, a negative feedback mechanism comes into operation, when individual meander loops become 'short-circuited' to form a **cut-off**. This process shortens the channel length, with a resultant increase in the channel slope. Two main types of cut-off are observed: **neck cut-offs** and **chute cut-offs** (Figure 8.14b). Neck cut-offs are the most common (Knighton, 1998) and several examples are seen in Colour Plate 11. The cut-off at the bottom right of this photograph would previously have been a double-headed meander bend.

Channel curvature is an important control on meander migration because of its influence on flow within the channel. Several researchers have found that meander migration is at a maximum for bends with an r_c/w

Plate 8.10 Valley confinement has restricted the lateral extension of these meander bends, resulting in their distorted shape, Maria's River, north central Montana, United States. Photograph by Airphoto, Jim Wark.

value of between 2 and 3. For example Hickin and Nanson (1975, 1984) reconstructed past rates of migration and channel curvature from former scroll bar deposits formed by the Beatton River in British Columbia, with maximum rates of meander migration occurring when the curvature was within this range.

Braided channels

Morphology and behaviour

Braided channels are very dynamic, with high rates of fluvial activity and rapid adjustments to channel form. Complex flow patterns are seen as the flow diverges and converges around the many bars. This results in large variations in bed shear stress across the channel, which influences patterns of scour and deposition. Major changes can take place at times of high flow, when rapid rates of channel migration are facilitated by high stream power and erodible banks. There can also be wholesale shifts in channel position as sub-channels are abandoned, or previously abandoned channels are reoccupied. The channel bars that typify braided channels are complex features. They are modified by processes of erosion, as well as deposition, and evolve over short periods of time. Various conditions are associated with braided channels. These are:

- *An abundant bedload.* This is supplied from upstream and enhanced by further contributions from bank erosion.
- *Erodible banks.* These allow a wide, shallow channel to form in which bars can develop. Channel subdivision continues until there is insufficient stream power to erode the banks (Leopold and Wolman, 1957).
- *High stream power.* Braided channels are often associated with steep slopes, although some large braided channels flow over low gradients.
- *A highly variable discharge.* Associated variations in flow competence lead to the sporadic movement of bedload, which is significant in bar development. However, braided channels can form under conditions of constant discharge (Leopold and Wolman, 1957; Ashmore, 1991). In the field, it is

not uncommon to find channels where braided reaches alternate with meandering ones. This suggests that a variable discharge may not be of primary importance.

The type of braiding and the number of sub-channels found within the main channel (braiding intensity) reflect the different environments in which braided channels are found. A number of **braiding indexes** have been developed, using different criteria, in an attempt to quantify the intensity of braiding. For instance, one type of index quantifies the number of active channels across the channel belt. Braiding indexes allow comparison between different channel reaches and can be used to assess channel changes over time. A distinction is often made between bars and islands, although these have the same origin and share similar morphological characteristics. While bars are only emergent at low flows, islands are more permanent, stable features, that may be vegetated and are only inundated by very high flows.

The three-dimensional shape of braided channels is not obvious from studying the planform alone, as this does not show variations in depth or the important links that exist between channels and bars. Individual sub-channels are typically curved in planform where they flow around bars. Field measurements have shown that the secondary flows which develop within these curved channel sections are similar to those in meander bends. Plate 8.11 shows a bar and the downstream confluence of two curved braid channels during low flow conditions. (The channel on the right hand side of the photograph is dry.) The photo is looking upstream. What you are looking at here is similar to two meander bends, back to back. The steep faces on each side of the bar have been formed by erosion on the outside of each bend, while finer sediment has been transported to the inside of the bends and deposited on more gently sloping bars.

The initiation and development of a braided planform

Various mechanisms are involved in the development of braided channel forms, depending on the channel

Plate 8.11 Confluence between two braid channels during low flow conditions. See text for explanation.

setting and controls. Bar formation is brought about by both depositional and erosional processes.

The mechanism of **central bar deposition** was first described by Leopold and Wolman in 1957 from laboratory flume experiments. This process is represented schematically by the diagrams at the top left of Figure 8.15, which incorporate the later observations of Ashmore (1991). In the top diagram, localised flow convergence at the upstream end of the reach leads to scour, with the eroded material forming a thin bedload sheet that migrates along the bed of the channel. Where flow becomes locally incompetent to transport the coarsest particles, they are deposited. These coarse deposits, called **lag deposits**, start to alter the flow pattern. This causes more sediment to be deposited, and leads to the upward growth and downstream extension of the incipient bar. As the bar expands, the flow is forced to flow around it, attacking the banks and widening the channel, which in turn produces more bed load.

In carrying out his own flume experiments, Ashmore (1991) found that a different process was dominant. He called this **transverse bar conversion** (see Figure 8.15). In the initial stages of this process, flow convergence leads to scour at the bed, forming a narrow, steep-sided **chute**. Erosion of the chute produces a large amount of sediment. Since the flow diverges as it exits the narrow chute, the flow competence rapidly declines and a lobe of sediment is deposited. This is an incipient bar. Successive bedload sheets stall (the bedload stops moving and is deposited) as they travel across the bar, building up the bar surface.

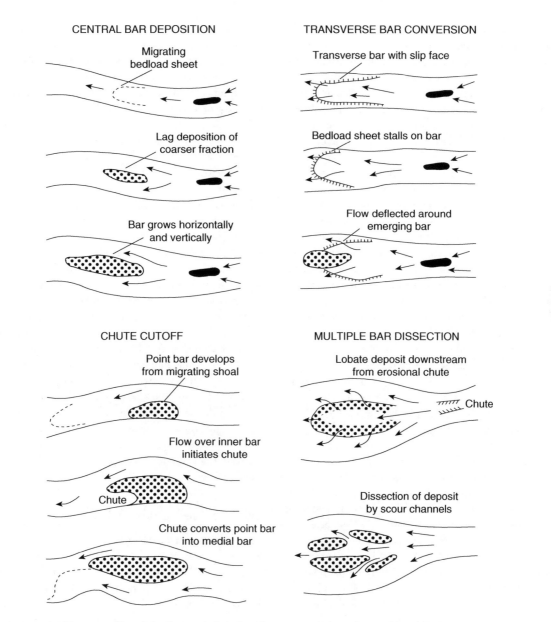

Figure 8.15 Mechanisms of braid development, based on flume experiments of Leopold and Wolman (1957) and Ashmore (1991) and cartoons from Ferguson (1993). After Knighton (1998).

As material accumulates, a steep slip face forms where the accumulated material starts to fall over the downstream edges of the bar. As the height of the bar increases, it starts to obstruct the channel, and flow becomes diverted around an elevated central lobe. This is different from the central bar mechanism, in that the initial bar deposits are formed by the erosion and wholesale deposition of large amounts of bedload, rather than the deposition of just the coarser material that the flow is locally incompetent to transport. The central bar mechanism appears to be more typical of situations where bed shear stress is only just above the threshold for movement, meaning that small variations in depth can render the flow locally incompetent to transport the coarser sediment. When the bed shear stress is considerably greater than that required to initiate bedload movement, it is possible for the large volumes of sediment required for the second mechanism to be eroded and deposited (Ashmore, 1991).

Bar development also involves erosional mechanisms, the first being the development of **chute cut-offs**. This involves the erosion of a channel (or chute) across a point bar, eventually separating it from the bank (Figure 8.15). Bars can also be dissected by multiple channels which form during high flow when the bar is submerged, resulting in **multiple dissection** (Figure 8.15). Bars are dissected during higher flow conditions, when water flows across the bar surface. Flow concentration leads to the formation of channels that cut into the bar surface. Dissected bars have the appearance, at higher stages, of being a number of smaller bars. Several examples of dissected bars can be seen in Colour Plate 6. You may also be able to identify bars that have been separated from the channel margins by chute cut-offs.

Avulsion is also important in braid development and describes a relatively sudden switching of the flow from one channel to another (Ferguson, 1993). This occurs when chute cut-offs form, or when the flow shifts to formerly abandoned channels. Other mechanisms include the blocking of channels by aggrading bars. This leads to ponding and the formation of an overflow channel.

Anabranching channels

There is some confusion surrounding the terminology of these multi-channelled forms, which are sometimes referred to as anastomosing. The nomenclature used by Nanson and Knighton (1996) will be used here, where the term **anabranching** is used to describe all planforms that are characterised by more than one separate channel. **Anastomosing** will be used to describe one particular subgroup of low-energy anabranching channels.

The flow in anabranching rivers is divided into two or more separate channels. Individual anabranches are typically incised into the floodplain, the resultant islands having a similar elevation to the surrounding floodplain. The islands are stable features which are often well vegetated and can sometimes remain relatively unchanged for decades or centuries (Knighton, 1998). Flow in each anabranch is largely independent of that of its neighbours (Bridge, 1993).

Although anabranching channels are fairly uncommon, they are the most diverse channel type and are found in many different environments. Examples come from climatic regions ranging from sub-arctic to tropical, and from monsoonal to semi arid (Knighton, 1998), forming in fine to coarse-grained alluvial substrates. Some anabranching channels are found in relatively high-energy environments while others have extremely low specific stream power. Despite this variety, there do appear to be some common characteristics, for example, these channels are usually characterised by flood-dominated regimes, and tend to have banks that are relatively resistant in comparison with the available stream power (Nanson and Huang, 1999).

Why do rivers anabranch?

Anabranching channels are often associated with very low slopes and, because little energy is available, the range of possible adjustments is somewhat limited. Although the slope cannot be increased, form adjustments can lead to a reduction in flow resistance, thus increasing the energy available for transporting sediment. It has been demonstrated that two or more channels with a low width-to-depth ratio (narrow and deep) are more hydraulically efficient than a single channel (Nanson and Huang, 1999). This is because the combined hydraulic radius of the multiple channels is greater (more hydraulically efficient) than for a single channel carrying the same flow.

Field observations show that individual anabranches tend to have a lower width-to-depth ratio than a single channel carrying the same flow. However, narrow channels can only form in cohesive or well vegetated alluvial substrates, otherwise instability and bank collapse lead to channel widening. If the individual anabranches are unable to maintain a low width-to-depth ratio, the hydraulic advantages of anabranching are lost.

Anabranching channels are usually formed by erosion, when avulsion leads to the incision of a new channel into the floodplain. Avulsion occurs during high flows, when one of the banks is breached and water spills out onto the floodplain. If flow is sufficiently concentrated, a new channel can be incised, eventually rejoining another channel further downstream. Some anabranches are only active during flood flows, acting as a distributory system for dispersing and storing water and sediment (Nanson and Huang, 1999). Individual anabranches are abandoned when they become infilled with sediment, perhaps as a result of a blockage or because the flow is diverted elsewhere. In some cases, anabranching may develop as a result of sediment deposition, when flow is concentrated by the development of bars or ridges within a relatively inefficient channel (Wende and Nanson, 1998).

Wandering gravel bed channels

At the high end of the anabranching energy spectrum are wandering gravel bed channels. These are intermediate between braided and meandering channels (Colour Plate 17). Wandering channels tend to develop where there are inputs of coarse sediment but where bedload transport rates are lower than for braided channels. Typical specific stream powers are within the range of 30–100 W m^{-2} (Church, 1983). Wandering channels are laterally active, although less so than braided channels, with fewer channels and active bars. The bars in wandering channels are more stable than those in braided channels, often being vegetated, and in most cases a dominant channel can be identified.

Anabranching behaviour in wandering channels is often associated with channel avulsion, where new channels are incised within the existing floodplain. This tends to happen at higher flows, perhaps as a result of a blockage, such as an accumulation of woody debris in one of the existing channels. The supply of coarse sediment also promotes the development of bars.

Anastomosing channels

The term anastomosing comes from medicine and refers to the branching and rejoining of blood vessels. This is rather an appropriate description for the rivers shown in Colour Plates 8 and 9.

Anastomosing channels are rare but can be found in a variety of environments. They share a number of common characteristics, which include low gradients, very low specific stream powers (less than 8 W m^{-2}) and stable banks formed from cohesive sediment or sand stabilised by riparian vegetation. In some cases the low-energy environment is caused by tectonic subsidence or an increase in the local or regional base level. Anastomosing channels are dominated by suspended sediment, and rates of bedload transport are generally very low. This combination of low energy availability and cohesive banks limits rates of lateral activity. The dominant process of floodplain formation is the vertical accretion of fine sediment.

Many examples can be found in Australia's two largest drainage basins: Lake Eyre and the Murray-Darling. Here, a combination of low gradients and an arid climate lead to a downstream decline in flow competence and a build-up of fine, cohesive sediment. Vegetation in the riparian zone is adapted to the arid climate, with deep root systems, further increasing bank stability. The mud-dominated anastomosing system of Cooper Creek in the semi-arid Channel Country of western Queensland is a well known example (Colour Plate 8). This has extensive anastomosed reaches along its length, which cover distances of over 400 km. The channel network is interesting in that there is a primary network of between one and four main channels with additional channels on the floodplain, which become active during flood flows (Knighton, 1998). Another feature of this system are numerous waterholes, that have developed along lines of preferential scour and bank erosion. The extreme flow regime for this river was discussed in Chapter 3 (see Figures 3.3b and 3.4).

Plate 8.12 Oblique aerial photograph of the Sandover–Bundey River near Ooratippra in the Northern Territory, Australia. Photography courtesy of Gerald Nanson.

The Columbia River (Colour Plate 9) is found in a rather different climatic setting, with a flow regime that is dominated by summer meltwater floods. Following the Rocky Mountain trench, the anastomosed reach is approximately 120 km long and changes to a wandering planform further downstream. The valley has a complex geomorphological history and was partly blocked by the growth of alluvial fans at the end of the last glaciation. This increase in local base level led to high rates of sedimentation on the up-valley side of the fans and a reduction in the valley slope (Tabata and Hickin, 2003). The channels of the present-day river are separated by levees. On the floodplain, low elevation wetlands are formed from vegetated silt and mud. During the annual summer floods, the basins between the channels become inundated, resulting in rapid rates of floodplain aggradation. Within the channels, coarse sand is transported as bedload, although the low stream power means that channel capacity is reduced by high rates of in-channel deposition.

Other types of anabranching channels

An interesting variant is found in the arid Alice Springs area of central Australia, where anabranching occurs in channels transporting coarse sand and gravel bedload (Plate 8.12). These are described by Tooth and Nanson (1999). The individual channels are separated by sub-parallel longitudinal ridges, stabilised by trees growing along the ridge crests. These ridges, together with wider vegetated islands, are formed either by processes of erosion, where they are cut out from the original floodplain, or by sediment accretion within the channel.

THE FORM OF BEDROCK AND MIXED BEDROCK-ALLUVIAL CHANNELS

In comparison with the extensive literature on alluvial channels, relatively little research has been conducted on bedrock-influenced channels. However, there has been a growing interest in these rivers over recent years.

Bedrock and alluvial channels exist along a continuum, and there are many examples of bedrock-alluvial channels where significant alluvial deposits reduce the extent of bedrock outcrops. In fact, it is not unusual to find several bedrock, bedrock-alluvial and alluvial reaches along the same channel. The underlying bedrock influences not only the channel substrate but also the valley slope and degree of valley confinement. Downstream changes in valley slope are brought about by variations in geology and lithology. This influences patterns of erosion and deposition, with erosion predominating in steeper reaches. Sediment deposition occurs along lower-gradient sections where sediment supply exceeds local transport capacities. While recognising this continuum, there are fundamental differences between its end members – channels formed entirely in alluvium and those formed directly in bedrock:

- *Channel adjustments.* Unlike alluvial channels, low to medium flow stages are usually relatively ineffective in shaping bedrock-dominated channels. It is true that a certain amount of abrasional wear and polishing is carried out at lower flows. However, the resistance of the substrate means that it is only high magnitude floods that are able to make significant morphological adjustments. Morphological features in bedrock channels, such as meanders and riffle–pool sequences, therefore tend to be scaled to higher flows. The comparative rarity of large events also means that time scales of adjustment are often much longer for bedrock channels. Bedrock-alluvial channels can potentially be adjusted to a very wide range of flows. This means that although the channel dimensions are scaled to major floods, depositional geomorphologic units within the channel are adjusted to much lower flows (Heritage *et al.*, 2001).

- *Dominant processes.* Deposition is limited along bedrock-dominated reaches. This means that channel development is mainly due to erosional processes. Bedrock channels are less able to adjust their form by deposition, reducing their degrees of freedom.

- *The influence of structural controls.* In some cases, structural controls imposed by the bedrock substrate exert the dominant influence on channel morphology. At the meso scale (centimetres to metres), channel form is influenced by joints, bedding and contacts between different lithologies. Initial weaknesses are preferentially eroded, eventually modifying the flow and leading to the development of irregularities in the channel. These include potholes, channel constrictions and 'forced' features such as bedrock steps or pools. Downstream changes in structural controls can occur over a short distance resulting in a number of short reaches, each with a different morphological character.

Bedrock and bedrock-alluvial channels display a wide variety of different channel forms. In planform, the flow usually follows a single channel, although there are also examples of anabranching bedrock channels. The differences in bed gradient and cross-sectional form discussed below are observed in all these types.

Bed gradient and channel cross-section

Bed gradients can be uniform or variable, with breaks in slope. Irregularities in the downstream bed gradient occur at the site of knickpoints (see below) and where there are sequences of steps and pools, or riffles, pools and rapids. Riffles and pools are observed in bedrock and bedrock-alluvial channels, with pools scoured to bedrock and accumulations of coarse sediment at riffles. These sequences may be initiated by channel constrictions. Although such constrictions are randomly spaced, the length of pools becomes scaled to high-magnitude flows, allowing a fairly regular sequence to develop (Thompson, 2001). Step–pool sequences are also observed in steeper reaches and can form entirely in bedrock.

Knickpoints are short sections of channel, along which the gradient is steeper than the adjoining reaches. Waterfalls are found where there is a vertical or near-vertical break in slope. Unlike the bedrock steps of step–pool sequences, knickpoints are not regularly spaced and are influenced by underlying structural controls. In tectonically active regions, knickpoints can be created by differential faulting or deformation. They are also associated with variations in the erodability of the bedrock, perhaps where the channel crosses bands of rock with differing resistance. The steep break in slope means that erosion is focused at knickpoints, which may become gradually flattened by erosion or retreat upstream.

Plane-bed channels (Plate 8.13) tend to form in densely jointed or relatively non-resistant rocks. This allows uniform rates of erosion, both across the channel and longitudinally, and results in a uniform channel with a relatively featureless bed (Wohl, 1998). While plane-bed channels often have a regular cross-section with a rectangular or trapezoidal shape, many bedrock channels are more irregularly shaped. Differential erosion across the channel results in features such as potholes, longitudinal grooves and inner channels within the main channel.

Bedrock channel planforms

Straight bedrock channels

Straight channels can broadly be divided into two types: those with straight walls and those with undulating walls. Undulating walls are often observed in deep, narrow slot canyons in both arid and humid environments. They are typical of channels that are actively incising through the upstream migration of knickpoints and potholes. Regularly spaced undulations form on alternate sides of the canyon and regulate energy expenditures along the reach (Colour Plate 13). Since the walls of a canyon form a much greater proportion of the channel boundary, it has been suggested that these 'wall forms' are analogous to bedforms (Wohl, 1998).

Straight-walled channels can have uniform or variable channel gradients and also exhibit a variety of cross-sectional forms. Variations in width are also common and along many channels, narrow, high-gradient reaches alternate with wider, low-gradient reaches that are covered with a veneer of sediment. These differences, which

often relate to variations in the underlying geological structures, are seen over a wide range of scales, from tens to hundreds of metres up to tens of kilometres. Large debris fans, deposited as tributaries join the main valley, can also affect channel width.

Meandering bedrock channels

Although many bedrock channels are straight or slightly sinuous, there are also examples of meandering channels. These are similar in form to their alluvial counterparts, although the flows that form them may have much longer recurrence intervals (Tinkler, 1971). Bedrock meanders were traditionally thought to have been inherited from a previous alluvial phase. Over thousands of years the meanders would be preserved, unchanged, as the channel incised into the underlying bedrock, perhaps as a result of tectonic uplift. This is true in some cases, where present-day flow and sediment regimes have little influence on meander sequences formed by catastrophic floods in the past. However it is now recognised that many bedrock rivers are actively meandering. When meandering rivers incise to create a gorge, the form of past meander sequences is preserved in the walls at different heights above the present channel. This indicates how meander loops have migrated and evolved during incision. You can see from Colour Plate 5 how flights of rock terraces have formed on the inside of meander bends as they have extended laterally. The narrow 'goosenecks' that separate two arms of a meander bend can also be breached, forming a cut-off. Well developed meanders are also observed in actively incising bedrock streams flowing through caves (Smart and Brown, 1981).

Evidence suggests that meander size can be greatly influenced by variations in substrate resistance, both as the channel migrates laterally as it incises vertically. Several studies have shown that meander dimensions are influenced by rock type. While resistant bedrock tends to preserve a meandering planform during incision, meanders tend to be smaller in softer bedrock, or decrease in size as the channel incises (Wohl, 1998).

Anabranching bedrock channels

Multiple bedrock channels can also be found. Preferential erosion along lines of weakness, such as joints and

Plate 8.13 Plane bed channel, Mpumalanga Province, South Africa.

fractures, can lead to the development of anabranching reaches. Tooth and McCarthy (2004) report examples from arid and humid regions including South and North America, India and South Africa. (A small-scale South African example is shown in Colour Plate 2.) This type of anabranching is also referred to in the literature as anastomosing or erosional braiding.

The aerial photograph in Plate 8.14 shows an anabranching bedrock reach of the semi-arid Sabie River, Mpumalanga Province, South Africa. Extensive bedrock outcrops are found along the Sabie, which has incised a wide macro-channel in which all but the most extreme floods are contained. Significant downstream variations are seen in channel morphology in response to changes in

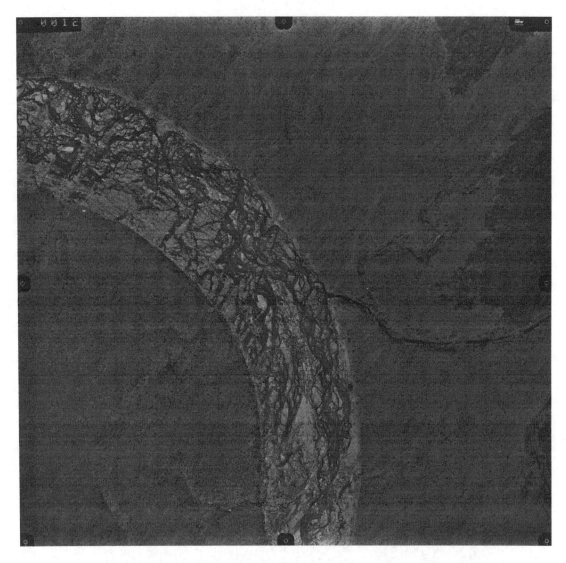

Plate 8.14 Aerial photograph of an anabranching bedrock reach of the Sabie River, Mpumalanga Province, South Africa. Photograph courtesy of South African National Parks and Centre for Water in the Environment, University of the Witwatersrand, Johannesburg.

the distribution and thickness of sediment deposits. The channel changes several times from a single to a multiple-channel form and various different bedrock, alluvial and mixed-reach types have been identified (Heritage *et al.*, 1999). Some of the anabranching reaches are characterised by extensive bedrock pavements, whereas along other reaches, deposition has created alluvial islands with bedrock core bars. Vegetation plays an important part in stabilising these deposits (Plate 8.15).

The aerial photograph was taken seven months after a major flood event, which took place in February 2000. This had an estimated return period of 200 years

Plate 8.15 A mixed anabranching reach of the Sabie River, South Africa. Bare sections of the bedrock bars can be seen in the centre of the photograph, which was taken from one side of the macro channel, the other side being marked by the distant line of trees.

(Heritage *et al.*, 2004). Given the extreme flood distribution associated with semi-arid regions, this was a huge event which resulted in major modifications to parts of the channel. Patterns of change were complex, with some deposits being largely unaffected by the flood. However, many of the mixed anastomosing reaches experienced widespread sediment stripping. The bedrock core bars and individual anabranches can be seen clearly in Plate 8.14.

CHAPTER SUMMARY

Channel adjustments are driven by the flow and sediment regimes and are the result of spatially-complex process–form feedbacks. There are potentially four main degrees of freedom for channel adjustment. These are channel cross-section, slope, planform and bed roughness. However, limitations are imposed by boundary conditions: the degree of valley confinement, the substrate in which the channel is formed, valley slope and riparian vegetation. Along a given river, downstream changes in the driving variables lead to systematic changes in discharge, channel slope, sediment size, channel dimensions and mean flow velocity. Since there are natural fluctuations in the supply of flow and sediment, channel form also varies over time. In the case of alluvial rivers, which tend to be adjusted to frequent, low-magnitude flows, fluctuations typically occur around an equilibrium channel form. However, not all channels are in equilibrium. Bedrock and ephemeral channels are typically adjusted to larger, but less frequent, flood flows. The form of alluvial channels and their associated floodplains varies along an energy gradient. This ranges from high-energy environments, where channels are dominated by coarse sediment, to low-energy, fine-grained channels. Many different channel forms exist along this continuum. Although 'typical' channel types can be identified, many channels do not fit neatly into categories, having characteristics of more than one type. Alluvial channels can broadly be divided into four main classes: straight, meandering, braided and anabranching. Straight and meandering reaches follow a single channel. Braided channels are characterised by numerous bars and islands, while the multiple channels of anabranching rivers are cut into the floodplain. Bedrock channels also show a wide diversity of forms but have received relatively little attention until recently.

FURTHER READING

More advanced texts

Brierley, G.J. and Fryirs, K.A., 2005. *Geomorphology and River Management: Applications of the River Styles Framework*. Blackwell, Oxford. Written from an Australian perspective, this well illustrated book includes details on the form and behaviour of a wide range of different channel types.

Knighton, D.A., 1998. *Fluvial Forms and Processes: A New Perspective*. Arnold, London. Develops the concepts introduced here with many illustrative examples from the literature.

Robert, A., 2003. *River Processes*. Arnold, London. Written to complement the book by Knighton, this has an emphasis on flow and sedimentary processes in alluvial channels.

Thorne, C.R., Hey, R.D. and Newson, M.D. (eds), *Applied Fluvial Geomorphology for River Engineering and Management*. John Wiley & Sons, Chichester, pp. 15–46. The chapters by Thorne, Hey and Hooke give further detail on channel types, morphology and adjustments.

Selected journal articles and research monographs

Ashmore, P.E., 1991. How do gravel bed rivers braid? *Canadian Journal of Earth Sciences*, 28: 326–41.

Includes some good photographs of braid bar development in a laboratory flume.

Best, J.L. and Bristow, C.S., 1993. *Braided Rivers*. Geological Society Special Publication No. 75. Geological Society, Bath. A collection of papers covering a very wide range of material including recent techniques and methodologies for the analysis of channel form. The paper by Ferguson is mentioned earlier in this chapter. The detailed paper by Bridge is also recommended, examining the interactions between channel geometry, flow, sediment transport and deposition.

Miller, J.R. and Gupta, A. (eds), 2003. *Varieties of Fluvial Form*. Wiley, Chichester, pp. 477–94. Includes a number of examples of rivers from more 'unusual' environments, which have previously received relatively little attention from researchers.

Nanson, G.C. and Croke, J.C., 1992. A genetic classification of floodplains. *Geomorphology*, 4: 459–86. Provides further detail on floodplain-forming processes and examines the influence of stream power and sediment character on floodplain development.

Tinkler, K.J. and Wohl, E.E. (eds), 1998. *Rivers over Rock: Fluvial Processes in Bedrock Channels*. Geophysical Monograph Series. American Geophysical Union, Washington DC. Contains chapters on various aspects of bedrock channel processes, form and behaviour.

Websites

Geomorphology from Space, http://eosdata.gsfc.nasa.gov/geomorphology. The online version of an out of print 1986 NASA publication which contains a gallery of space imagery, location maps and commentary. There is an excellent chapter on fluvial geomorphology where you will find more information on some of the examples used in this book.

Earth Science World Image Bank (American Geological Institute), http://www.earthscienceworld.org/images/index.html. A searchable collection of images which contains a number of different fluvial landforms. A short explanation accompanies each image.

9

SYSTEM RESPONSE TO CHANGE

The fluvial system is affected by changes in the external basin controls. These affect the channel controls, in turn leading to changes in channel form. In this chapter you will learn:

- What causes change.
- How the fluvial system responds.
- Why some rivers are more sensitive than others.
- How past changes are reconstructed.
- How river channels have been affected by climate, human activity, tectonics and base-level changes.

THE NATURE OF CHANGE

In the previous chapter you saw how the form of a channel is controlled by driving variables (flow and sediment supply), and boundary conditions (valley slope and confinement, channel substrate and riparian vegetation). Also considered were the ways in which these variables control the adjustments that can be made by a given channel.

At any given point in time, these controls are themselves influenced by the past history of the drainage basin. For instance, the nature of an alluvial channel substrate is determined by former flow and sediment regimes. Similarly, valley slope is adjusted over time during phases of incision and aggradation. The sediment yields of formerly glaciated basins in British Columbia are still influenced by the vast quantities of sediment that were produced by glacial erosion (Church and Slaymaker, 1989). Fluvial systems therefore retain memories of past events, some of which go back a long way. The long-term influence of past events is sometimes referred to as a historical hangover (Ferguson, 1981).

Several examples that illustrate the effects of inherited controls have been mentioned in previous chapters. The low-gradient anastomosing reach of the Columbia River is a legacy of the complex geomorphological history of the valley (Chapter 8, p. 150). In California, the Sacramento River and its tributaries are still affected by huge volumes of sediment produced by gold mining in the nineteenth century (Box 5.3).

When considering present conditions, the flow and sediment regimes are generally assumed to be relatively constant. Over longer time scales this is no longer true. Changes in the external basin controls – primarily climate, tectonics and human activity – affect flow generation and sediment production within the drainage basin. This frequently results in modifications to channel form. These can be difficult to predict, although it may be possible to predict the direction of change, such as an increase or decrease in channel width (Box 9.1). Channel changes that are triggered by the external

Box 9.1

PREDICTING THE RESPONSE TO CHANGES IN DISCHARGE AND BEDLOAD SUPPLY

Channel form adjusts in response to changes in discharge and bedload supply. For example, an increase in bedload supply typically leads to an increase in the width to depth ratio (channels become wider and shallower). Simple algorithms can be used to indicate the direction of channel change. These were derived by Schumm (1969) on the basis of observations made for rivers in semi-arid and sub-humid regions. Most of these were sand-bed channels.

Discharge (**Q**) and bedload supply (**Qb**) can increase (+), decrease (−) or remain unchanged. The resulting changes in form are indicated + for an increase, − for a decrease and ± where change could be in either direction.

An increase in discharge (Q+) is typically associated with increases in channel width (**w**), depth (**d**), width to depth ratio (**w/d**), meander wavelength (λ) and a decrease in channel slope (s). With a decrease in discharge (Q−), these effects are reversed:

$$Q+ \longrightarrow w+ \quad d+ \quad (w/d)+ \quad \lambda+ \quad s-$$

$$Q- \longrightarrow w- \quad d- \quad (w/d)- \quad \lambda- \quad s+$$

The effects of increases and decreases in bedload supply have the following effects, and also influence sinuosity (**S**):

$$Qb+ \longrightarrow w+ \quad d- \quad (w/d)+ \quad \lambda+ \quad s+ \quad S-$$

$$Qb- \longrightarrow w- \quad d+ \quad (w/d)- \quad \lambda- \quad s- \quad S+$$

In practice, the basin-scale response to changes in climate and land use usually affects both the flow regime and the sediment regime. There are four possible scenarios:

Discharge and bedload supply increase together. This occurs during the expansion of urban areas.

Large quantities of sediment are generated during construction, and surface permeability is reduced.

$$Q+ \quad Qb+ \longrightarrow w+ \quad d\pm \quad (w/d)+ \quad \lambda+ \quad s\pm \quad S-$$

Discharge and bedload supply both decrease. An example is when improved land management practices are implemented to reduce soil erosion. Improved basin management also involves the re-instatement of wetland areas on the floodplains of engineered rivers. This has the effect of reducing flood peaks.

$$Q- \quad Qb- \longrightarrow w- \quad d\pm \quad (w/d)- \quad \lambda- \quad s\pm \quad S+$$

Discharge increases and bedload decreases. This might occur as a result of increasing humidity in an initially sub-humid area (e.g. savannah or steppe grasslands). Increases in discharge would be accompanied by an increase in vegetation cover, reducing rates of sediment supply.

$$Q+ \quad Qb- \longrightarrow w\pm \quad d+ \quad (w/d)\pm \quad \lambda\pm \quad s- \quad S+$$

Discharge decreases and bedload increases. Such a situation arises when mining activity takes place within a drainage basin.

$$Q- \quad Qb+ \longrightarrow w\pm \quad d- \quad (w/d)\pm \quad \lambda\pm \quad s+ \quad S-$$

These simple algorithms indicate no more than the potential direction of change. They are not capable of predicting the extent of change, or the rate at which it will take place. Another problem is that the form components do not all adjust at the same rate. The components that are easiest to adjust will change more rapidly (Chapter 8, pp. 128–129). In addition, the equations on which the algorithms are based were derived for a limited range of channels and climatic conditions. As such they may not be universally applicable outside this range.

controls are called **allogenic changes**. If you refer back to the flow diagram shown in Figure 2.2, you will see that shifts in the external controls can have direct and indirect effects on river channels. For instance, a change in climate will have a direct influence on the volume of runoff. However, vegetation growth will also be affected. This will, in turn, influence processes of runoff generation and sediment production. In addition, the susceptibility of the channel to bank erosion will be altered by changes in the growth of riparian vegetation.

The form of a channel can change without there having been any external disturbance. Examples of these **autogenic changes** include the development of meander cut-offs, various types of avulsion, and the formation and erosion of braid bars. When an internal threshold is crossed, autogenic changes can be dramatic. For instance, episodes of vertical accretion and dramatic floodplain stripping are observed for channels in southeast Australia (Nanson, 1986). Over periods of hundreds to thousands of years, floodplains are built up by vertical accretion of fine overbank sediment. Then a single large flood creates catastrophic erosion, stripping the floodplain surface down to its coarse basal deposits. Following this, the floodplain is slowly built up again. During floodplain accretion, the growth of large levees progressively displaces overbank flow from the broad floodplains and into the channels. This concentrates stream power and erosional energy. Eventually, a point is reached when the energy of a large flood is sufficiently focused to strip the floodplain sediment. Although a number of similar floods will occur during phases of floodplain accretion, it is only when the system is close to the threshold that floodplain stripping takes place.

Types of disturbance

Disturbances, and the effects that they have, occur over a wide range of space and time scales. Landslides and localised faulting are examples of local-scale disturbances. A wider area is affected by floods and regional tectonic uplift, while glaciation affects whole continents. Over the long term – thousands to tens of thousands of years or more – changes have been driven mainly by shifts in climate, tectonics and variations in base level. In the medium term – the last few thousands

of years – human activity has had an increasing influence. Indeed, human impacts have tended to dominate in the more recent past.

External disturbances to the system can be described as being either pulsed or ramped disturbances (Brunsden and Thornes, 1979). A **pulsed disturbance** is an episodic event such as an extreme flood, which is sudden and short-lived. The effects of pulsed disturbances are localised, and are usually followed by a return to the original state. If the system is close to a threshold, recovery might not occur. In this case, lasting changes are sustained by a shift towards a new equilibrium state.

Ramped disturbances represent a long-term shift in one or more of the controlling variables. Examples include an increase in the frequency of large floods, gradual tectonic subsidence, or urban development.

System response

Reaction/relaxation times

There is usually a lag between the disturbance and the associated response. This is illustrated in Figure 9.1. The top two graphs illustrate the two different types of disturbance that might occur: pulsed (top graph) or ramped (middle graph). The lower graph shows the response of the system. Prior to disruption, the system is in a state of equilibrium. Once the disturbance has occurred, there is a **reaction time** over which the impact of the change is absorbed by the system. This is followed by a period of adjustment, the **relaxation time**, as the system moves towards a new equilibrium. Eventually the system becomes adjusted to the new condition, and a new equilibrium is reached. In the case of a pulsed disturbance this adjustment could represent a return to the original state.

There are huge variations in response rates. In some cases an instantaneous response is seen, while in others, adjustment can take thousands of years. A number of different factors have to be taken into consideration. These include the type of disturbance, the scale at which it occurs, the ability of the system to adjust and the sensitivity of the system to change. Added to this are the effects of complex response (Chapter 2, pp. 16–17), as the effects of change are propagated through the system.

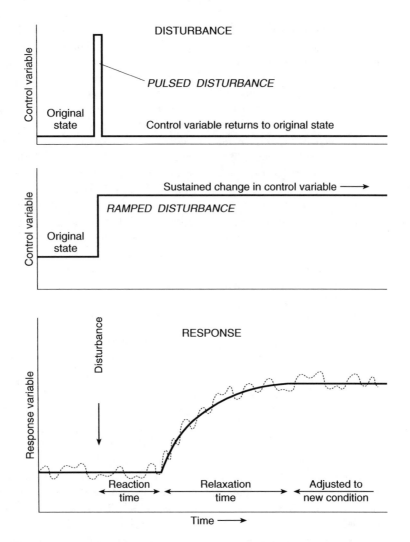

Figure 9.1 Pulsed (top) and ramped (middle) disturbances. The lowest graph illustrates how the system responds to a disturbance. The dotted line represents observed values, the solid line indicates the mean condition of the system. Adapted from Bull (1991).

Internal feedbacks can accentuate the effects of a disturbance. This occurs when knock-on effects lead to positive feedbacks, pushing another part of the system over a threshold. Conversely, negative feedback mechanisms can act to buffer and reduce the impacts of external changes.

Different parts of the system respond at different rates. An increase in precipitation will have an almost instantaneous effect on stream discharge. Associated changes in vegetation cover take longer, with research suggesting a response time of about 100 years (Knighton, 1998). Channel form components, such as meander wavelength, reach gradient and profile form, adjust over longer time scales (Figure 8.4).

Sensitivity to change

There are large variations in the sensitivity of different river systems to change. For example, a meandering channel that is close to the meandering–braided threshold may well shift to a braided form if there is an increase in stream power. A channel that is not close to this threshold will be less sensitive.

A useful way to understand sensitivity is to examine the balance between the driving forces that lead to change and the resisting forces that oppose it. A shift in this balance will result in change. For example, incision might be caused by the increase in stream power (driving force), associated with more frequent floods. A change in resistance might be brought about by the removal of riparian vegetation. This made the Slesse Creek in British Columbia more sensitive to erosion, leading to channel widening (Plates 8.3 and 8.4).

The **capacity for change** is determined by the degrees of freedom that a given channel has (Chapter 8, p. 129). The immediate effects of a large flood will be rather different for a sediment-starved channel, flowing through a bedrock gorge, than for a sand-bed meandering channel bordered by a wide floodplain. Since the resistance of the bedrock channel is high, it will show little change in comparison with the sand-bed channel. However, a truly catastrophic flood would be capable of making significant modifications to the bedrock channel.

Now consider the **recovery potential** of the two channels from the catastrophic flood. For the purposes of this illustration, it will be assumed that both channels are in a humid, temperate region. The sand-bed channel would have a high recovery potential. This is because temperate alluvial channels are shaped by relatively frequent low-magnitude events (Box 3.3). Of course, if this channel was close to a threshold, the flood might lead to a shift towards a new equilibrium and long-term change. In comparison, the recovery potential of the bedrock channel would be much lower. In fact, the relaxation time might be so long that the channel does not fully recover before the next rare and devastating flood. Bedrock channels are supply-limited. This means

that the capacity for making 'depositional adjustments', such as channel narrowing, aggradation and bar development, are greatly reduced. Recovery potential is also reduced in transport-limited environments, which are influenced by large influxes of sediment. Examples include rivers that have been directly affected by glacial sediment or mining activity.

Something called **event sequencing** is also important. The effects of a given storm event might not be particularly spectacular. However, if the same storm was preceded by a larger event, it might be rather more effective. For example, the soils and regolith on the steep slopes of the headwaters might have become highly saturated. Further rainfall could just tip the balance, leading to a number of failures. It has been demonstrated that floods are more effective in making channel adjustments if they have been preceded by a larger event (Beven, 1981).

Response to a given storm event can also be spatially variable across river systems. For example, Chapter 8 (pp. 154–155) referred to an extreme flood that occurred along the bedrock-influenced Sabie River in South Africa in 2000. Observations made along a 108 km length of channel showed that widespread sedimentary stripping had occurred along some reaches, while others remained essentially unchanged (Heritage *et al.*, 2004).

RECONSTRUCTING PAST CHANGES

Field-based approaches

Direct observation

Several different types of evidence are used to understand how rivers change through time. For time periods of less than a few decades, direct observations can be made. These include monitoring flow and sediment transport, surveys of channel form and measurements of bank erosion rates. Short-term changes can be assessed on the basis of continuous monitoring, or by making repeat measurements at intervals. Depending on the scale and type of observation, monitoring periods range from seconds to years. For instance, detailed measurements of small-scale flow and sediment dynamics are made over very short periods of time. At a larger spatial scale, rates of bank erosion would need to be assessed over several years. Direct observations can be very detailed, but only provide a brief 'snapshot' of a system that evolves over much longer periods of time. For example, the data may capture a small fluctuation on a general trend. This could lead to erroneous conclusions about the behaviour of the system.

Records of precipitation and discharge often go back over several decades, much longer in some cases. The first continuous programme of discharge measurements was started by the US Geological Survey for American rivers in 1889, while intermittent records for the Nile go back over 1,000 years. Unfortunately, the monitoring of sediment loads, especially bedload, is much more limited. Like many observations, these measurements are more likely to be carried out for a specific research project rather than as part of a long-term monitoring programme.

Surveying techniques are often used to measure channel form parameters. For smaller channels, direct field surveys are often more appropriate than using aerial photography, or satellite remote sensing (see below). Increasingly sophisticated equipment and techniques have been developed and include electronic distance measuring (EDM) theodolites and GPS surveys. Using repeat surveys, changes in channel position, cross-sectional dimensions and slope can be monitored. Surveys can be very time-consuming, and are rarely carried out for more than one or two decades. This is a limited window of time, and means that the range of events may not be fully representative in the longer term.

Historical records

Historical records provide various sources of information about past channel changes. These include documents, survey notes, maps and photographs. Using these records, changes in channel size, morphology and position can be reconstructed. In doing this, it is important to consider the accuracy of early maps and surveys. Existing discharge records can be extended back using flood marks on bridges and buildings. Other information on past floods can be obtained from newspaper reports and personal accounts.

Aerial photography and satellite remote sensing

Aerial photographs provide a valuable tool for assessing change. Past channel changes are evident from some of the aerial photographs in this book. These are indicated by meander scars, cutoffs and palaeochannels. In some cases, a sequence of aerial photographs may be available for an area. Gilvear *et al.* (2000) used measurements from aerial photographs to assess rates of bank erosion on the meandering Luangwa River, Zambia, between 1956 and 1988. Unfortunately, only a very small percentage of the land surface is covered in this way.

Small-scale aerial photographs, which cover a relatively wide area, are useful for examining large rivers and drainage networks. Photographs taken at lower elevations cover a smaller area in more detail, allowing changes along individual reaches to be examined. Near-infrared aerial photography is particularly useful for assessing vegetation and variations in moisture content. Using photogrammetry, it is also possible to make three-dimensional measurements of surface topography from stereoscopic pairs of photographs. These are vertical views of the same scene, slightly offset from each other. High resolution topographic data can be obtained using aircraft scanning laser altimetry. This allows detailed mapping of floodplain topography (see below).

There have been major advances in satellite remote sensing since the first Landsat satellite was launched in 1972. Today there are a number of Earth observation satellite platforms, each carrying several sensors. These sample the electromagnetic radiation that is reflected and emitted from the Earth's surface.

The frequency of imaging and resolution are determined by the orbit of a given satellite. Like a photograph from a digital camera, satellite images are composed of a number of pixels. One edge of a pixel represents a length on the ground surface. This length increases with the height of the orbit, ranging from a few metres to a kilometre or more. Pixel size limits the size of fluvial landforms that can be 'seen' by a given sensor. Anything smaller than a pixel will be difficult to identify and impossible to delineate. In practice, a number of pixels are needed to identify a given feature (Gilvear and Bryant, 2003). This means that the greater spatial resolution provided by aerial photographs is more appropriate for smaller rivers or for detailed reach-scale surveys. However, satellite remote sensing is invaluable for mapping large rivers in remote locations. The high frequency of imaging means that changes can be monitored over time. The potential also exists for monitoring variations in suspended sediment concentrations over extensive reaches of channel, although this is at an early stage of development. It should be noted that field-based measurements are needed to support and interpret all types of remotely sensed data, including aerial photographs.

Airborne and terrestrial LIDAR

Airborne LIDAR (light detection and ranging) enables laser-based measurements to be made of the distance between a sensor, carried by an aircraft, and the ground surface. Data collection is rapid, and the technique has considerable potential in generating high-resolution surveys of complex channel topography. In applying the technique to the gravel-bed River Coquet in Northumberland, England, Charlton et al. (2003) found a close correspondence between ground survey-derived cross-profiles and those generated by LIDAR. They noted that some correction may be needed for anomalies in the LIDAR surface, which are associated with areas of deep water and vegetation. It can also be difficult to determine the precision of LIDAR measurements for any one survey.

Some limitations are imposed by the resolution of airborne LIDAR data, meaning that data at the grain and microtopographic scale are not provided (Heritage and Hetherington, 2007). However, developments in oblique ground-based laser scanning offer improved accuracy, resolution and aerial coverage. For example, Heritage and Hetherington applied oblique field-based laser scanning techniques to produce digital elevation data, to a resolution of 0.01 m, for an upland reach of the River Wharfe, in Yorkshire, England. This is a dynamic channel, with a rapidly changing morphology, varied topography and vegetation cover. While the errors reported were relatively small, the accuracy of such data is dependent on a number of factors. These include the field operation of equipment, the nature of

the terrain, and the instruments themselves. Heritage and Hetherington (2007) suggest a field protocol for obtaining optimal results.

Sophisticated surveying techniques, such as laser altimetry, can be integrated with digital elevation models (DEMs). Spatial analysis software is now highly advanced, allowing complex analyses to be made. This makes it possible to measure and monitor morphological change at a range of scales. Lane et al. (2003) demonstrated the potential of these powerful new tools in estimating volumes of erosion and deposition in the braided Waimakariri River, New Zealand. Using digital photogrammetry, laser altimetry and image processing, DEMs were constructed for a 1 km × 3.3 km area of river bed. The data were collected in February–March 1999 and February 2000. DEMs of difference (between the two years) were constructed by subtracting DEM pairs. These had a pixel resolution of 1.0 m, and realistic patterns of erosion and deposition could be identified. Volumetric estimates of change were also made. These estimates represent an improvement on those derived from existing cross-sectional surveying techniques. Although traditional techniques provide higher precision for a given point, this is more than compensated for by the much higher density of spatial data provided by the new techniques (Lane et al., 2003).

Fluvial records

A record of past changes in the fluvial environment is preserved in the sediments stored on floodplains and terraces. Careful analysis of these deposits can provide insight into past channel behaviour and changes that have occurred. The record is complex and incomplete, because sediment is reworked by lateral migration. Part of the record is also destroyed during phases of floodplain erosion. When channel incision takes place, remnants of former floodplain surfaces are often preserved as **terraces** at a higher elevation above the present-day floodplain (Box 9.2). Floodplains can also be partly abandoned when the magnitude and frequency of floods decreases and inundation becomes a rarer event. This allows mature soil profiles to develop, which may be preserved as **palaeosoils** within the sedimentary record.

Box 9.2

TERRACE FORMATION

Incision into the floodplain surface leads to floodplain abandonment and the formation of river terraces. These are relatively flat-topped benches found at the edge of valleys. A steep scarp slope separates the original floodplain surface from the present-day floodplain below. Stair-like sequences of terraces can also form, with each 'step' representing a former floodplain surface (Colour Plate 19). Terraces are partly destroyed when the modern river erodes into them, as it migrates across the floodplain. This means that terraces may be preserved only as fragments.

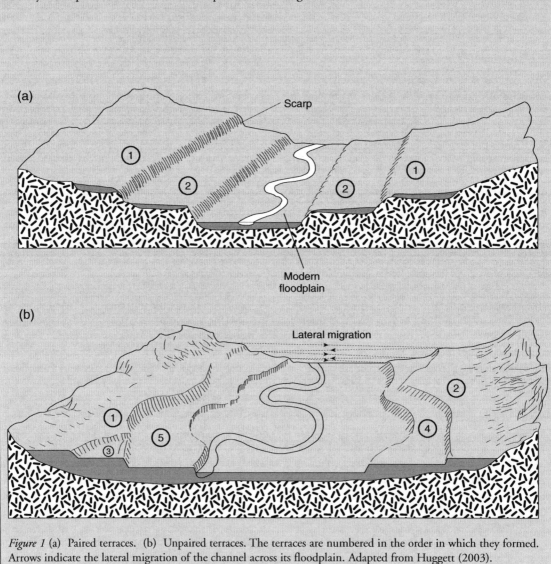

Figure 1 (a) Paired terraces. (b) Unpaired terraces. The terraces are numbered in the order in which they formed. Arrows indicate the lateral migration of the channel across its floodplain. Adapted from Huggett (2003).

Box 9.2

TERRACE FORMATION—CONT'D

Terraces can be formed when the channel incises into alluvial valley fills, or into bedrock. Bedrock terraces are usually covered with a thin veneer of alluvial sediment, and are sometimes called **strath terraces**. Like the modern valley floor, terraces slope gently in a downstream direction. However, the steepness of this slope is often different from the present floodplain because of changes in energy conditions and sediment supply.

When incision is rapid, with rates of downcutting exceeding lateral migration, **paired terraces** are formed. These are found at the same elevation, on opposite sides of the valley (Figure 1a). More common are the **unpaired terraces** that are created as an incising channel migrates across its floodplain (Colour plate 19). Figure 1(b) shows five terraces that have formed sequentially in this way. Terrace 1 is the oldest in this sequence, after which the incising channel migrated across the floodplain. Terrace 2 was then formed on the opposite side of the valley, at a lower elevation. Subsequent migration has created the later terraces in the sequence. You will see that much of terrace 3 has been destroyed by subsequent erosion.

Incision and terrace formation is often caused by a fall in the local, or regional, base level. This can be brought about by tectonic or isostatic uplift, or a fall in sea level. An increase in stream power, or a reduction in the supply of bedload, can also lead to incision. The balance between stream power and sediment supply was discussed in Chapter 8 (see Figure 8.3). Changes in this balance can precipitate phases of degradation and terrace formation, or aggradation and valley fill. These changes can take place without there having been any external influence, as part of normal cycles of adjustment.

Palaeochannels can be found on the floodplain surface, and may be fully, or partly, infilled with sediment (Colour Plate 18). Below the surface are older channels, which are buried beneath more recent sediments. Palaeochannel cross-sections are often exposed when the present-day river erodes and migrates. Associated with these former channels is an assemblage of channel and floodplain features, such as bars, lateral accretion deposits and bedforms. Careful analysis of these structures can provide information on the conditions that existed when they were laid down. This allows former flow and sediment regimes to be reconstructed. It is also possible to determine how much energy was available, and whether the channel was dominated by bedload or suspended load.

Using relationships derived for modern rivers, the dimensions of former channels can be used to estimate former flow conditions. Bankfull discharge can, for example, be established from measurements of meander wavelength. An alternative method uses measurements of cross-sectional area and channel slope. Structures, such as bedforms, provide further clues, indicating flow direction and velocity. Analysis of particle size distributions is also used to estimate bed shear stress and velocity.

A number of chronologically distinct, morphological and sedimentary assemblages may be contained within the fluvial record. Changes in channel morphology and behaviour, whether gradual or abrupt, can be observed within this record. Inferences can then be made about the changing environmental conditions that drove these changes. The first stage in analysis is to identify separate assemblages of deposits. During phases where continuous aggradation occurred, successive deposits are laid down above older deposits. These can then be dated according to the order in which they were deposited. This is called **relative dating**. Pollen and archaeological artefacts preserved within the deposits are used to aid interpretation. Unfortunately, such sequences are often rather complex. Sedimentary records are frequently destroyed during periods of incision or floodplain stripping. During later phases of alluviation, new deposits are laid unconformably on this erosion surface. This creates an abrupt transition, representing a gap in the record. To complicate things further, the lateral migration of channels reworks former deposits and replaces them with new deposits.

In order to build up a 'sequence of events' it is therefore necessary to use some kind of dating technique. To find out the actual date before present (BP) **absolute dating** techniques are used. In the past, dating techniques using radioactive isotopes have been used. These include **Carbon-14**, **Caesium-137** and **Lead-210** dating. **Dendrochronology** uses tree rings to provide dates over the last 2,000 years, and was used by Hickin and Nanson (1975, 1984) to date meander scrolls on the Beatton River, in order to determine rates of migration (Chapter 8, p. 145). One of the most promising techniques for dating Holocene sediments is **optical dating**. This has been used to date various fluvial deposits, including floodplain sediments, palaeochannels and terraces (Rodnight et al., 2006). Optical dating provides an estimate of the time that has elapsed since sand and silt-size particles were last exposed to solar radiation. An approximate date of burial by deposition can be determined by exposing particles to a beam of light and measuring the intensity of emitted light (Bell and Walker, 2005). Buried quartz grains are exposed to weak fluxes of ionising radiation. Within the crystal structure of these grains, electrons are detached (ionised) from their parent nuclei. These electrons then become 'trapped' within the crystal lattice, the number of trapped electrons increasing with time since burial (Bell and Walker, 2005). Exposure to sunlight empties the electron 'traps' and resets the time 'clock'. This is commonly referred to as 'bleaching'. Problems are associated with 'partial bleaching', where particles are not exposed to sufficient sunlight to empty all the electron traps. During fluvial transport, grains follow different pathways within the water column. Individual grains are therefore exposed to varying strengths of sunlight for different lengths of time. This exposure may not be of sufficient strength to fully remove the signal from the previous burial period. Deposited grains may therefore carry a residual signal, or 'memory' of an earlier period of burial. This can lead to over-estimation of the length of the most recent burial period. Careful sampling procedures, analysis and interpretation are therefore necessary (Rodnight et al., 2006).

In comparison with alluvial channels, the fluvial records for bedrock channels are rather more limited. Nevertheless, it is sometimes possible to reconstruct past flood events using **slackwater deposits**. During floods, slackwater areas are formed by flow separation, allowing fine sediment to settle out of suspension. These deposits are found on ledges, alcoves and in bedrock caves, and mark the peak stage of palaeofloods. O'Connor et al. (1994) reconstructed fifteen major floods on the Colorado, using slackwater deposits to extend the record back over 4,500 years.

Recent developments in **cosmogenic exposure age dating** now mean that rates of bedrock incision can be determined. This estimates the length of time that rocks have been exposed at the surface. For example, Schaller et al. (2005) used cosmogenic dating to determine rates of incision within a marble gorge in the active mountain belt of Taiwan. Dating of fluvially sculpted surfaces on the gorge walls revealed exposure ages ranging from 200 BP at the base of the gorge, to 6500 BP at a height of 165 m above the present channel.

Computer simulation

Since the 1980s there have been major advances in computer simulation models of fluvial processes at many different scales. Such models provide a valuable tool for interpreting and understanding the complexities of change. At one end of the scale are high-resolution **computational fluid dynamics models (CFD)**, which are applied at the channel reach, or sub-reach scale. These are based on the fundamental physics of flow. Simulations represent short periods of time, such as a single flood event, and flow calculations are made at very small time intervals. These models have been used in many different applications, and there have been recent developments in the simulation of bank erosion and channel adjustment (see Pizzuto, 2003, for a review). A significant development is in the application of **cellular models**, which are considerably less demanding in terms of computer resources. These can be used to provide a two-dimensional representation of braided rivers, alluvial fans and even whole drainage basins (see below) by a mesh of grid cells. The transfer of water and sediment between cells is simulated using simple rules, which are based on the underlying physics that governs these processes (Nicholas, 2005).

Landscape evolution models simulate slope and channel network development over thousands, or even millions, of years. The nature of drainage basin response to external changes, such as climate and human activity, is highly complex. Interpretation of sequences of events from the sedimentary record can be very difficult, particularly for larger drainage basins. Since the Last Glacial Maximum, fluvial systems worldwide have been affected by major variations in climate. Widespread changes in land use have also occurred, altering the response of drainage basins to climatic variations. Added to this are the effects of complex response within the drainage basin (see Chapter 2, pp. 16–17). As a result, the links between external changes and internal response are far from simple. For example, alluviation may be delayed, with different responses occurring both within and between drainage basins (Richards, 2002).

Computer modelling provides a means of understanding and assessing basin response to external change. Modellers are able to control a simulated environment, for example by altering precipitation inputs and modifying land cover. Comparisons can then be made between different responses. Most landscape evolution models represent the landscape (for example, a drainage basin) using a cellular modelling approach. Water is routed across a mesh of grid cells representing the basin, and the elevations of individual cells are changed to represent fluvial and slope erosion (Coulthard, 2001). Increases in computing power and data availability have allowed larger basins to be simulated, with improved representation of channel and hillslope processes (e.g. Willgoose *et al.*, 1991; Howard, 1994). Models are now starting to be applied to real drainage basins, rather than theoretical ones. For example, Coulthard *et al.* (2005) applied a landscape evolution model to four upland catchments, all tributaries of the Yorkshire Ouse, England. Responses to fluctuations in precipitation and land cover over the Holocene (the last 10,000 years) were simulated. The model simulations showed that, although identical environmental changes were applied, there were significant differences

in coarse sediment transfer and alluviation between basins. The model allowed sediment volumes to be tracked downstream, making it possible to analyse the response of individual downstream reaches. Within the different basins, the sensitivity of different reaches varied considerably. For example, some periods of alluviation were only recorded in certain reaches. Coulthard *et al.* suggest that these differential responses may be related to the passage of sediment waves, and spatial variations in sediment storage and availability. Also significant are large- and small-scale thresholds for sediment transfer within each basin. A comparison between model output and the actual record of Holocene river erosion and alluviation showed some correspondence. However, it is important to note that these simulations represent a simplification of the environmental record, as well as the processes of sediment transfer (Coulthard *et al.*, 2005). The authors point out that more field evidence is needed, both to validate some of the simulated responses produced by the model and to increase model sophistication.

Any model represents an abstraction of the actual system that is being modelled. Models simplify reality, with only those components that are perceived to be significant being represented in a given model (Wainwright and Mulligan, 2004). As Wainwright and Mulligan point out, models aid in understanding but are not an alternative to observation.

RESPONSE OF FLUVIAL SYSTEMS TO CHANGE

The role of extreme floods

Extreme floods are pulse disturbances that are capable of carrying out large amounts of geomorphological work in a short period of time. The effectiveness of large floods depends on a number of factors. These were outlined in the earlier section that looked at sensitivity to change.

The immediate effects of a given flood are determined by the balance between force and resistance to change. The force, in terms of specific stream power and shear stress exerted, can reach very high values where the flow is concentrated. For example, the Narmada River in central India flows through a narrow gorge at Punasa.

The monsoonal flood regime of this river is highly variable, with frequent large flood events. During monsoonal floods, the flow depth varies from 13 m in wider bedrock sections to 60 m where it is confined to the gorge. Associated specific stream powers range from 2,600 W m^{-2} to 12,800 W m^{-2} and shear stresses from 300 N m^{-2} to 3,000 N m^{-2} (Rajguru *et al.*, 1995). Field observations have shown that these higher stream powers are capable of eroding bedrock, eroding and transporting boulders as bedload and carrying cobbles in suspension. After a catastrophic flood on the sand-bed Wollombi Brook in New South Wales the amount of change was found to vary in direct proportion to the degree of valley confinement (Erskine, 1996). The duration of floods is also significant in determining the volume of sediment moved and the extent of channel modifications. On the other side of the force–resistance balance is the susceptibility to erosion. For instance, arid zone rivers are particularly vulnerable to erosion, because of sparse vegetation and low bank resistance.

Recovery times vary greatly – indeed some channels may not 'recover' at all, moving instead towards a new equilibrium form. A frequently cited example of **delayed recovery** is the Cimarron River in Kansas United States, which was widened from 15 m to 365 m by a series of major floods. Prior to this, a number of dry years had resulted in a decline in riparian vegetation, which greatly reduced bank resistance. The increased width was maintained, until a series of relatively wet years allowed vegetation to become re-established. This allowed the channel to regain its original width (Schumm and Lichty, 1963).

While humid alluvial channels often show relatively little change, or recover rapidly, bedrock gorges in arid environments preserve the effects of large, catastrophic floods for long periods of time (Baker and Pickup, 1987). When the time between major floods is shorter than the recovery time, the channel is more likely to have a flood-dominated morphology.

Late Quaternary climatic change

Over the last 125,000 years the global climate has been characterised by glacial–interglacial cycles caused by variations in the Earth's orbit. For most of this time,

glacial conditions prevailed, with intervening interglacials occurring only 10 per cent of the time. These huge changes in climate have had major impacts on river systems worldwide.

The most recent period of glaciation reached its maximum 18,000 years before the present (abbreviated as 18 ka BP). Since then, the climate has undergone a transition from glacial to the present post glacial conditions. During this major climatic transition, three main periods of fluvial activity can be identified (Knox, 1995):

- The period of the Last Glacial Maximum, from 20 ka to 14 ka BP.
- The period of the major climatic transition, from glacial to post glacial conditions, from 14 to 9 ka BP.
- The period during which modern natural vegetation patterns and ocean–atmosphere circulation regimes became established, from 9 ka BP to the present.

Fluvial responses to these changes in climate are recorded worldwide, even in areas that were not directly affected by glaciation. This is because global atmospheric circulation patterns were affected by the development of extensive ice sheets. Positive and negative feedbacks within the ocean–atmosphere system caused displacement of the climate belts, with a resultant reorganisation of prevailing winds and storm tracks.

The following discussion represents a highly simplified overview of patterns of change. It is important to realise that there were considerable global variations in the timing and nature of climatic change. In addition to climatic influences, many drainage basins were also affected by tectonics and sea level change (discussed later in this chapter). Further variation has resulted from the modification of land use by human activity.

20 ka to 14 ka BP

Eighteen thousand years ago, huge ice sheets covered much of North and South America, Europe and Asia. At lower latitudes, mountain glaciers expanded and advanced. Beyond the maximum extent of the ice sheets vast areas were affected by periglacial conditions.

Extensive areas at lower latitudes were much drier than they are today.

Rivers draining the ice sheets were supplied with huge volumes of meltwater and sediment. The sediment yield was further increased by a greatly reduced vegetation cover. Vast braided systems flowed over rapidly aggrading outwash plains. Active aggradation along the braided upper Mississippi and its tributaries created a longitudinal profile that was much steeper than it is today. This can be seen from the height of the former river, preserved as a terrace, which lies 45 m above the modern floodplain. A further 1,000 km downstream the same terrace lies only 5 m above the present floodplain (Knox, 1995).

In present-day tropical and subtropical regions, conditions of aridity existed over large areas. This led to increased aeolian activity, with dune fields covering extensive parts of Africa, India, Australia and North and South America. These aeolian deposits are still significant, affecting many rivers in these regions today (Macklin and Lewin, 1997). In Ghana and Sierra Leone (West Africa), rain forest was replaced by savanna grassland between about 21 ka and 13.5 ka BP. During this period of aridity, low rates of fluvial activity were observed, with infrequent flooding and limited evidence of channel migration (Thorp and Thomas, 1992).

14 ka to 9 ka BP

This second period marked the transition from glacial to postglacial conditions. As the ice sheets started to melt, rates of coarse sediment supply were initially very high. However, improvements in the climate led to the re-establishment of vegetation, which reduced sediment loads. The large proglacial lakes, which formed in front of retreating glaciers, also reduced sediment supply, acting as highly efficient sediment traps. Former periglacial regions became invaded by forest vegetation.

Although there was a decline in the sediment supply, high discharges were maintained by large volumes of meltwater and increased levels of precipitation. This resulted in a change from an aggradational to an erosional regime, and many temperate zone rivers underwent a change from braiding to meandering. The timing of this transition was dependent on a number of factors, such as the timing of glacial retreat and the volume of sediment that was stored in the valley system during the Last Glacial Maximum. However, it tended to occur at the same time as vegetation became established (Starkel, 1991b).

Associated with this period are large palaeomeanders, preserved in former floodplain deposits. These have a much longer wavelength than the meanders of modern rivers that occupy the same valleys today. An example can be seen in Figure 9.2, which shows large palaeomeanders in the valley of the Prosna, a tributary of the Warta, on the Polish Plain. It is estimated that the discharge that formed these meanders was between 3.6 to 6.5 times the modern discharge (Rotnicki, 1991). The traces of discontinuous braided channels predate the meandering channel.

During the transition from glacial to post glacial conditions the ice sheets waxed and waned several times, in response to fluctuating temperatures. During colder episodes, mountain glaciers re-advanced and many rivers reverted to a braided form. The last of these cold phases was the **Younger Dryas**, which lasted from 12.9 ka to 11.6 ka BP. The effects of this are best known in Europe, although there is evidence to suggest that conditions were cooler and drier at lower latitudes (Knox, 1995).

Also associated with this period are the catastrophic floods that occurred when vast proglacial lakes overflowed. There are several well known examples associated with the melting of the Laurentide ice sheet, which covered much of North America. These include Glacial Lake Missoula, a meltwater lake that formed at the end of the last glaciation. On several occasions, enormous volumes of water were released from this lake. The flow was so great that it exceeded the capacity of the preflood valleys and multiple channels were carved as water spilled over drainage divides (Baker, 1973). These catastrophic floods formed the area known as the Scablands in eastern Washington, United States.

There is also evidence in the fluvial record for increased precipitation in subtropical and tropical latitudes. In tropical West Africa, Thorp and Thomas (1992) identified a period of higher discharges. This is thought to represent a return to wetter conditions, with high flood peaks resulting from large rainstorms.

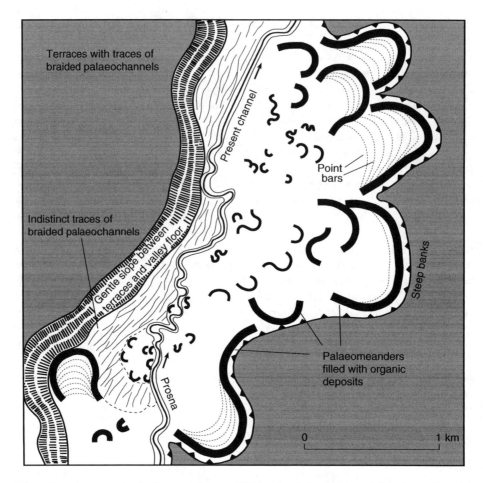

Figure 9.2 Palaeochannel patterns in the Prosna valley near Wieruszów, Poland. Adapted from Kozarski and Rotnicki (1977).

Knox (1995) cites evidence from the Niger River delta, where much higher discharges led to increased rates of sedimentation between 12.2 ka and 10.5 ka BP. Also, in East Africa, increased rainfall occurred on the Lake Plateau between 12 ka and 10 ka BP. This is thought to have been associated with a period of higher discharges, leading to degradation along the Nubian Nile (Said, 1993).

9 ka BP to the present

The period from approximately 10 ka BP is called the **Holocene**. This has been a period of relative stability, in comparison with the large climatic shifts that occurred

during the transition from glacial to interglacial conditions. During the early Holocene there was a considerable reduction in the discharge of most rivers in the temperate zone. For example, Dury (1977) suggested a sevenfold decrease in the mean river discharge, and a threefold decrease in sediment loads. It is also widely recognised that rivers changed from being bedload-dominated, through mixed-load, to suspended-load channels. This led to the development of smaller meanders, with a marked reduction in meander wavelength.

From 9 ka to 4 ka BP, rapid warming occurred. The warmest period was around 5 ka BP, when temperatures were 1–2°C higher than they are today. This was followed by an episode of climatic deterioration, with cooler and wetter conditions from about 3 ka to 2.5 ka BP. Even moderate fluctuations in temperature and precipitation can lead to significant changes in the intensity of fluvial processes. Large shifts in flood frequency–magnitude relationships are associated with relatively small changes in temperature. A change of just 1–2°C can result in changes in precipitation of up to 10–20 per cent (Knox, 1995). In mid-latitude regions, cooler episodes are associated with more frequent large floods. Regional correlations can be seen, with intensified fluvial activity in Europe associated with periods of glacial advance (Starkel, 1991b). Macklin and Lewin (1993) examined stratigraphic records from a number of sites in Britain and identified several major phases of valley alluviation. These are synchronous with similar events in the United States and Europe (Knighton, 1998). The most recent phase identified by Macklin and Lewin coincides with the **Little Ice Age**, a period of climatic deterioration occurring between AD 1550 and AD 1750. This followed the **Medieval Warm Period**, during which temperatures were warmer than they are today. This lasted from AD 900 to AD 1250.

During the later part of the Holocene, human activity has had an increasing influence on river systems. Deforestation and development of land, first for grazing and, later, cultivation, has changed flow and sediment regimes. As such, it can be very difficult to distinguish the effects of human activity from those caused by climatic fluctuations.

In tropical West Africa forest had become re-established in Ghana and Sierra Leone by about 9 ka BP. Peak discharges appear to have remained high and frequent, with rivers scouring down to bedrock. There is also evidence for major reworking of floodplain sediments. A period of floodplain stability appears to have occurred between 7 ka and 4.5 ka BP. However, after approximately 4.5 ka BP, flood peaks increased, leading to overbank flows and floodplain construction (Thorp and Thomas, 1992).

Short-term climatic variability

Over time scales of decades, cyclical shifts in the frequency of flooding have been identified for rivers in various locations. These are caused by non-random variations in rainfall that are associated with upper atmospheric circulation patterns. In the coastal region of New South Wales, Australia, rainfall-driven variations cause the flow regime to oscillate, between a flood-dominated regime (FDR) and a drought-dominated regime (DDR), every thirty to fifty years (Erskine and Warner, 1988, 1999). When FDR conditions prevail, there is an upward shift in the frequency–magnitude relationship (pp. 31–34). This means that extreme flood events become more frequent, leading to channel widening and the erosion of in-channel benches. During periods dominated by DDRs the frequency of large floods decreases, leading to relatively long periods of fluvial stability. Channels contract as benches are re-formed. High frequency changes in climate and flood regime have also been identified in the Tyne basin in northern England, where alternating phases of incision and stability have occurred since AD 1700 (Rumsby and Macklin, 1994). Short-term climate change effects can be very difficult to isolate from the effects of other changes.

Human activity

Changes within the drainage basin

Deforestation increases the sensitivity of soils to erosive forces, accelerating erosion and sediment delivery to the channel network. Removal of the topsoil also exposes the relatively impermeable lower soil layers. This reduces infiltration rates, enhancing storm runoff and increasing the magnitude of flood peaks. In general, runoff and sediment yields are relatively high from cultivated and intensively grazed land, when compared with forested areas.

Evidence of human activity can be traced back over thousands of years. Phases of valley floor sedimentation are associated with periods of forest clearance and agricultural expansion (Chapter 5, p. 65). In Europe, many forest regions were cleared for cultivation as long ago as 6 ka BP. In some steppe regions, runoff and sediment yields were increased by human activity 9.5 ka BP (Starkel, 1987).

Sensitivity to land use change varies from place to place and is dependent on factors that include soil erodability, climate, drainage basin topography and slope–channel coupling. The Yellow River in China has one of the highest sediment yields in the world. Much of this sediment comes from the cultivated land of the

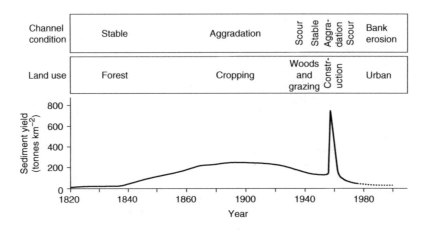

Figure 9.3 Sequence of development of channels in urban areas. Adapted from Wolman (1967).

highly erodible Loess Plateau (Figure 5.4). The sediment yield would have been much lower before 200 BC, when the plateau was mainly unfarmed wooded steppe. Milliman *et al.* (1987) suggest that, by AD 600, the plateau was farmed heavily, resulting in a tenfold increase in the sediment load of the Yellow River.

Sediment overloading of rivers results in broader, shallower and less sinuous channels. Many Central European rivers underwent a change, from meandering to braided, in the seventeenth and eighteenth centuries. This was due to extensive agriculture, the introduction of potato plantations and the higher flood frequencies of the Little Ice Age (Starkel, 1991a). Vertical accretion rates can be accelerated by an order of magnitude or more. For example, deforestation in the Drury Creek drainage basin, southern Illinois, United States, led to the deposition of at least 2 m of fine sediment on valley floors. The maximum period of settlement and deforestation was from the 1860s to the 1920s. During this time, rates of valley floor deposition have been accelerated, by an estimated one to two orders of magnitude. Since the 1940s, sediment yields have been reduced by the introduction of soil conservation measures (Miller *et al.*, 1993).

The growth of urban areas increases runoff and sediment yields, although the effects tend to be more localised. Figure 9.3 shows the effects of land use change on sediment yields for the Piedmont Region of the United States. Forest clearance increased sediment yields, which eventually reached a plateau. Later, the urbanisation of agricultural areas led to a sharp increase in sediment yield during the construction phase. Following construction, sediment yields declined. Further reductions were associated with the introduction of soil conservation measures (Wolman, 1967). The large increase in sediment yield is typical of expanding urban areas. In the Denver area of Colorado, United States, increased rates of sedimentation on floodplains were associated with urban expansion (Graf, 1975).

Although there is a decline in sediment supply following construction, the flow regime is permanently altered. Runoff is accelerated by impermeable surfaces, such as paved areas, roads and car parks. Drains and sewers also act as efficient conduits, rapidly transmitting water to river channels. Depending on how much of the drainage basin is urbanised, the size of flood peaks can be greatly increased. This effect is greatest for smaller, more frequent floods. Hollis (1975) showed that, for a 20 per cent urbanised basin, a tenfold increase is seen in the one-year event. However, the one-in-two year event only increases by a factor of 2 to 3. Channel enlargement may be associated with these increases. In a study of British urban sites, the bankfull cross-sectional area was found to enlarge by a factor of 1.61 on average. In some cases, increases of up to six times were observed (Roberts, 1989).

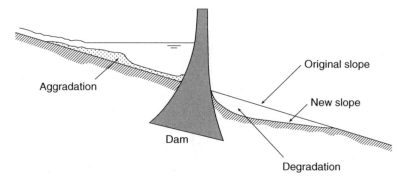

Figure 9.4 Aggradation and degradation at a dam. After Gordon *et al.* (2004).

In areas affected by mining, dramatic changes may be associated with the production of huge volumes of mine waste (see Chapter 5, pp. 65–67).

Direct modifications of the channel

The construction and operation of dams alters both flow and sediment regimes. Flow regulation often reduces flood peaks, and modifies the seasonal distribution of flows. Dams also act as barriers to sediment transport (Chapter 5, p. 67), which in turn affects channel form. Upstream, where the river flows into the reservoir, the loss of competence means that all the bedload is deposited. Over time, this leads to the build up of a wedge of sediment (Figure 9.4). In some cases the effects of this are fairly localised, although a wave of aggradation may slowly propagate upstream (Leopold and Bull, 1979). Most of the fine material carried in suspension is trapped when it settles out in the still waters of the reservoir. In large reservoirs these losses can account for up to 99 per cent of the suspended load (Williams and Wolman, 1984).

The sediment-starved flow downstream from dams is aptly described as 'hungry water'. During high flow releases, the absence of a sediment load results in an energy excess. If this exceeds the resistance of the channel substrate, bed erosion and channel incision will occur (Figure 9.4). Downstream, incision is most effective on fine-grained substrates and where reductions in flood peaks are relatively small. However, little or no degradation

has occurred on many regulated gravel bed rivers in upland Britain, because most flow releases are below the transport threshold for the coarse bed material (Petts and Lewin, 1979). Incision can sometimes affect extensive lengths of channel. In their study of twenty-one US dams, Williams and Wolman (1984) observed downstream migration of degradation at rates of up to 47 km a year. Changes in planform may also occur. The Stony Creek, a tributary of the Sacramento, in California, originally flowed in a braided channel, but since closure of Black Butte Dam, in 1963, this has changed to an incised, meandering channel (Kondolf and Swanson, 1993).

Various negative feedbacks prevent incision from proceeding indefinitely. For example, stream power is decreased when the channel slope is reduced by incision. Selective removal of finer fractions of the bed material may lead to the development of a protective armour layer (pp. 103–04). Another limit is imposed when outcrops of bedrock are exposed by erosion. These act as local base-level controls on upstream reaches, where the channel bed cannot erode below the height of the outcrop.

There can be a reduction in channel capacity if flood peaks are seriously reduced by flow regulation. This is often brought about by the growth of fine-grained bars and berms at the channel margin. Sediment is supplied to the main channel by unregulated tributaries, which join it further downstream from the dam. As a result, the greatest decreases in width are often observed below tributaries (Knighton, 1998). Reductions in bankfull cross-sectional area of more than 50 per cent have been

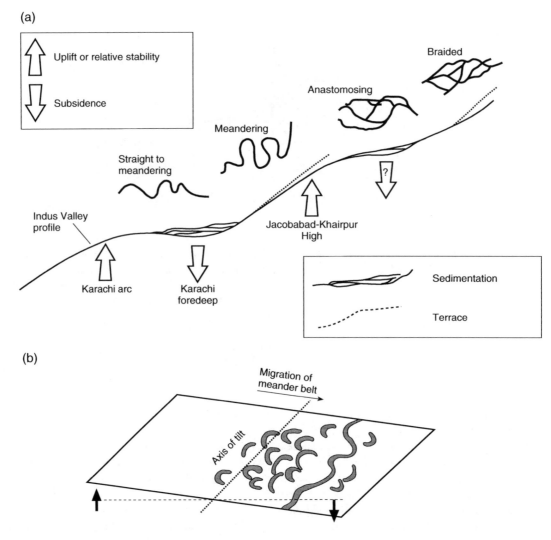

Figure 9.5 (a) Schematic diagram showing the relation of the Indus valley profile and channel pattern to tectonic elements. The slope and pattern changes are exaggerated for illustration. After Harbor *et al.* (1994). (b) Meander belt migration as a result of lateral (cross-valley) tilting.

observed (Petts, 1979). Narrowing can also be aided by the encroachment of vegetation into the channel. This would formerly have been flushed out by flood flows.

Patterns of adjustment downstream from dams are complex. Controls on adjustment include the extent and type of regulation, bed material size, and the type of sediment brought in by unregulated tributaries further downstream. While channel adjustments can take place very rapidly, changes to sediment transfer regimes occur over time scales of 10 to 500 years (Petts, 1984).

Channelisation is the modification of river channels, for purposes of flood control, navigation, land drainage and erosion control. Channel engineering techniques include straightening, channel enlargement, erosion control and the building of embankments. In addition to the obvious effects at the site, instabilities are often created upstream and downstream from the engineered reach. The impacts of channelisation will be discussed more fully in Chapter 10.

Tectonics and base-level change

Tectonics

Throughout geological history, the Earth's crust has been deformed by internal forces. The resultant warping, folding, faulting and tilting have had a profound effect on river systems worldwide. Tectonically active regions, such as the Himalayas and Andes, have some of the highest sediment yields in the world. Long-term, large-scale uplift, subsidence or deformation disrupts channel networks and long profiles. At the reach scale, channel adjustments are associated with changes in slope, lateral tilting and localised faulting.

The effect of tectonics on slope and channel form is illustrated by the lower Indus as it flows across the alluvial plain of Sindh. The Indus, which rises in the Himalayas, has a high sediment load and aggradation predominates in its lower course. The structural elements underlying the lower Indus valley are undergoing different types of tectonic movement, which include uplift, tilting and subsidence. The resulting changes in slope, together with varying rates of sedimentation, are the primary influence on the channel pattern (Harbor *et al.*, 1994). The resulting downstream variations in channel form are represented schematically in Figure 9.5(a). Anastomosing reaches are associated with a rise in the local base-level caused by uplift immediately downstream. Downstream from areas of uplift, the Indus meanders again, the degree of sinuosity increasing with the valley slope. Where the channel actually crosses the crest of areas of upward deformation, incision has led to the development of terraces (Jorgensen *et al.*, 1993).

Lateral tilting results in floodplain surfaces that are higher on one side of the valley than the other. The axis of tilt runs parallel to the line of the valley (Figure 9.5b). Instantaneous uplift can create large displacements of several metres. This may lead to the immediate avulsion of rivers towards the side of the floodplain that has subsided. A different response, where tilting is more gradual, is a preferential migration of the meander belt across the floodplain. The meander scars that are left behind mostly

face in the same direction (Alexander *et al.*, 1994). This is because only those that form on the higher side of the floodplain will be preserved (Figure 9.5b).

Even in 'stable' regions, away from active plate margins, slow warping of large areas of the Earth's crust can lead to significant deformation over time. Burnett and Schumm (1983) observed significant variations in channel morphology for rivers crossing two uplift domes in the American states of Louisiana and Mississippi.

Base-level change

Base-level change can occur at a local or regional scale. Causes include tectonic uplift or subsidence, changes in sea level and isostatic uplift of land masses. Local base-level changes can be brought about when aggradation or incision in the main channel alters the base level of the tributaries that join it. As mentioned earlier, human activity, such as dam construction, also affects local base levels.

During the Last Glacial Maximum, global sea levels were 130 m lower than they are today. Because more land was exposed, many drainage basins covered a larger area than they do now. River courses were longer too, particularly where shallow continental shelves lie beyond the present day coastline. The fall in sea level increased the potential energy of the landscape, often leading to incision. However, the effects of this were usually fairly localised and limited to the lower reaches. In some cases, active incision of the continental shelf created deep valleys. These exist today as submarine canyons. Much of the increase in sea level, to the present level, had taken place by about 6 ka BP. This submerged the lower course of many rivers. Associated reductions in potential energy led to downstream deposition, as channels adjusted their gradients.

Sea-level changes are referred to as **eustatic** changes. The influence of eustatic changes on base levels is complicated by the **isostatic** effects of land movement. During the glacial period, considerable pressure was exerted by the weight of ice on the affected continents. After the ice melted, the removal of pressure resulted in an upward rebound of continents. This isostatic uplift involves long lag times and still continues today. Some of the highest rates of isostatic uplift – up to 2 cm a year – are observed in west central Sweden. Isostatic uplift can reduce, or even reverse, the effect of sea-level rise in some cases. Changes in the relative rates of eustatic sea-level rise and isostatic uplift have resulted in wide variations in base-level change from place to place.

CHAPTER SUMMARY

Changes in the external controlling variables – climate, human activity, tectonics and base level – influence the whole fluvial system. As a result of internal feedbacks, the response of river channels to such change is complex and often involves significant time lags. Channel change can also occur without there having been any change in the external basin controls. This is called autogenic change and is brought about when internal thresholds are crossed. Disturbances of the fluvial system, or part of it, can be pulsed – where a single event, such as a major flood, takes place – or ramped – where there is a long-term shift in one of the controlling variables. The response depends on the nature and scale of the disturbance, and the ability of the system to adjust. Evidence of change comes from a number of sources, including direct observations, historical records, maps, remote sensing and sedimentological evidence. Fluvial records, preserved as floodplain deposits, are often very complex, necessitating the use of dating techniques to build up a sequence of events. Since the Last Glacial Maximum there have been major shifts in global climate, which have affected river systems worldwide. Associated changes in flow and sediment yields have led to phases of alluviation and degradation. There have also been dramatic changes in channel pattern and behaviour. Human activity has had an increasing influence on fluvial systems. This started several thousand years ago, with forest clearance for grazing and, subsequently, cultivation. More recently, the effects of urbanisation and direct channel modifications, such as dam construction and channelisation, have had major impacts on river systems. Some channel networks and long profiles are disrupted by tectonic effects, which include warping, tilting, folding and faulting. Reach-scale modifications are also seen. The lower reaches of many rivers have been affected by the rise in sea level since the last glaciation. These effects vary considerably, being complicated by isostatic uplift of the land surface. Local changes in base level, brought about by various mechanisms, are also significant.

FURTHER READING

More advanced texts

Brierley, G.J. and Fryirs, K.A., 2005. *Geomorphology and River Management: Applications of the River Styles Framework*. Blackwell, Oxford. The chapters 'River change' and 'Geomorphic responses of rivers' to human disturbance give clear explanations and also include sections on assessing vulnerability to change.

Brown, A.G., 1996 Floodplain palaeoenvironments. In: M.G. Anderson, D.E. Walling and P.D. Bates (eds) *Floodplain Processes*. John Wiley & Sons, Chichester, pp. 95–138. Provides more information on floodplain stratigraphy and floodplain evolution during the Holocene in north-west Europe.

Knighton, D.A., 1998. *Fluvial Forms and Processes: A New Perspective*. Arnold, London. The final chapter provides a more detailed discussion of channel change, with numerous examples.

Knox, J.C., 1995. Fluvial systems since 20,000 years BP. In: K.J. Gregory and L. Starkel (eds), *Global Continental Palaeohydrology*. John Wiley & Sons, Chichester, pp. 87–108. This chapter examines climatic and human influences on river systems since the last glacial maximum.

Schumm, S.A., 2000. *Active Tectonics and Alluvial Rivers*. Cambridge University Press, Cambridge. If you are interested in the influence of tectonics on alluvial river channels this is well worth a look.

Techniques

Benito, G., Lang, M., Barriendos, M., Llasat, M.C., Francés, F., Ouarda, T., Thorndycraft, V., Enzel, Y., Bardossy, A., Coeur, D. and Bobée, B., 2004. Use of systematic, palaeoflood and historical data for the improvement of flood risk estimation: review of scientific methods. *Natural Hazards*, 31: 623–43. An accessible account of reconstructing floods from palaeoflood and historical data.

Kondolf, G.M. and Piégay, H. (eds), 2003. *Tools in Fluvial Geomorphology*. John Wiley & Sons, Chichester. A very useful reference on field and modelling techniques which provides much more detail on all aspects of analysing channel change.

10

MANAGING RIVER CHANNELS

Industrial, agricultural and urban development place ever-growing pressures on land and water resources. Traditional management priorities have met human needs, with little regard for ecosystems. On a global scale the resulting impacts on the physical, chemical and ecological condition of river systems have been truly profound. There are many examples of severe environmental degradation, and few rivers can be described as being in a pristine or near-pristine condition. However, over the last part of the twentieth century, growing environmental awareness led to a shift in management priorities. This was coupled with developments in the understanding of river behaviour, dynamics and change. The role of the river manager now includes new challenges, where environmental considerations must be integrated with the development of water resources and the management of hazards, such as flooding. River management is now multidisciplinary, involving experts from a number of different fields, including geomorphology, ecology, engineering and economics. Geomorphologists have an important role to play in assessing the condition of rivers and their catchments, their sensitivity and future response to change. In this chapter you will learn about:

• The ways in which river channels have traditionally been managed to meet human needs.

• The adverse impacts of past land and water management on the physical, chemical and ecological condition of channels and floodplains.
• Changes in management philosophy towards more environmental approaches.
• Environmentally sensitive management techniques.
• Restoration of degraded channel reaches.

TRADITIONAL ENGINEERING TECHNIQUES

Why rivers are engineered

Human settlements have long been located along river channels, which provide a supply of water and power, fertile floodplain soils, fisheries and a potential means of navigation. Rivers can also be hazardous and many urban areas are increasingly at risk from flooding as they expand onto floodplains. This risk is further increased by larger flood peaks associated with land use change within the drainage basin – upstream deforestation, land drainage and urban development can all significantly increase flood peaks further downstream. Flood control works involve artificially increasing the channel cross-section, constructing flood embankments, straightening channels and removing vegetation and other obstacles. More recent techniques include the construction of flood diversion channels and flood storage reservoirs.

Many lowland rivers are maintained for navigation; these include the Rhine, Danube, Mississippi, Missouri, Ohio and Arkansas. The aim is to maintain a minimum depth of water along the navigable length of the river by means of dredging, removal of shoals and other obstacles and river training (see below). Weirs and locks are also used to extend the navigable length, providing a minimum depth for larger vessels at all times, despite natural variations in discharge.

Increased industrialisation and urbanisation place growing demands on water supply systems. The world's largest cities consume water at a rate that is exceeded only by the flow of a few major rivers. Meeting these demands involves constructing dams to store and regulate flow. It is also necessary to integrate supply from a number of different surface and subsurface sources, which can involve transferring water over large distances in river channels, pipelines and canals. On a global scale, the largest demand comes from irrigated agriculture in arid and semi-arid environments. Advances in irrigation technology allowed a huge expansion in the total irrigated area during the Green Revolution of the 1960s. This was also the decade that saw the construction of the greatest number of dams, the scale of which had been increasing since the construction of the Hoover Dam in 1935. By 1986 there were 39,000 large dams over 15 m in height (International Commission on Large Dams (ICOLD), 1988) and today few of the world's major rivers are unregulated. In the second part of the twentieth century there was a growing trend towards multi-purpose dams, whose roles include water supply, flood control and hydroelectric power generation.

Channel modifications are often involved in land reclamation and the drainage of wetlands and low lying areas. Many channels in lowland areas have been deepened and straightened to convey the increased volume of water resulting from the installation of field drains. Local modifications are also made to channels where channel instability might cause problems, for example, to prevent bank erosion at the site of bridges and other structures.

Channelisation and flow regulation

Channelisation is the modification of natural river channels for the purposes of navigation, flood control, land drainage and erosion control (Brookes, 1988).

Re-sectioning and realignment

Re-sectioning describes the modification of the channel cross-section to provide adequate depth for navigation and to increase the channel capacity for land drainage and flood control. This may involve the removal of a few bars or shoals, or deepening all, or part, of the cross-section. Some channels may be enlarged further by widening. Depending on the size of the channel and the purpose of the engineering works, re-sectioning is carried out using dredging, or by means of river training.

Realignment involves the straightening of river channels for purposes of navigation and flood control. It is also carried out where channels share the valley with roads and railways, to reduce the number of bridges that have to be constructed. In navigable rivers with a high natural sinuosity, the removal of meanders greatly reduces the distance that has to be travelled by vessels moving up and downstream. However, the increase in gradient can lead to instability as a result of increased stream power in the straightened section. Increased erosion in this section can lead to problems of deposition further downstream.

Dredging

Dredging is the removal of sediment from the bed of the channel for flood control and to maintain or deepen existing navigation channels. It is also carried out when sand and gravel are mined from the river bed. Dredging has been carried out for thousands of years and was practiced by the Egyptians, Sumerians, Chinese and Romans, by means of mass labour and manual tools (Petersen, 1986). Over time, developments in dredging technology and mechanisation have enabled these operations to be carried out at increasing scales.

In smaller, non-navigable channels, dredging is carried out from the bank using a dragline or bulldozer. Bank-side vegetation is often removed to enable access. In larger channels, the dredger is mounted on a floating platform. Mechanical dredgers remove material by lifting it from the bed in a bucket or dipper, whereas hydraulic dredgers use suction pumps to remove material, via a pipeline, to a disposal site. Rotating cutter heads and explosives are used to remove resistant bed sediment and bedrock outcrops.

Dredging for channel maintenance has to be carried out on a regular basis, at considerable cost. This is because it treats the problem (sediment accumulation) rather than its causes (sediment sources). Excessive mining of sediment from the river bed can lead to serious problems of erosion, both upstream and downstream from the site (p. 67).

Snagging and clearing

Flow resistance in the channel is increased by woody debris (fallen trees, logs, branches), large rocks and urban debris. In-channel debris presents additional hazards to navigation and may threaten bridges and other structures. The purpose of snagging and clearing is to remove this material, and these operations are usually part of the routine maintenance carried out every few years on many engineered channels. Trees and bushes at the edge of the channel may also be cleared at the same time. This is because bank-side vegetation can increase resistance to high flows, reducing velocity and potentially increasing the flood risk. Other reasons for removing vegetation are to allow access and to reduce the amount of woody debris entering the channel. As with dredging, snagging has to be carried out on an ongoing basis.

Levees and embankments

Levees are artificial embankments which are built alongside or close to the channel margins of lowland rivers. Their purpose is to increase the channel capacity at high flows and protect the surrounding floodplain from inundation. Levees are found extensively along many major rivers, including the Nile and Mississippi. Traditionally, levees have been constructed of earth, and many still are. In urban areas, where the potential human and economic losses are greater, levees and floodwalls are usually made of concrete. It is not feasible in economic or practical terms to construct levees that would contain all the floods that could possibly occur. Levees are therefore built to withstand a certain **design flood**, such as the twenty-year event. If this flow is exceeded the levees will be overtopped. The depth of flow contained within the levees is greater than it would be if no levees were present and water was able to inundate the floodplain. Since shear stress increases with flow depth (see Box 6.1), increased erosion of the channel bed is possible.

Bank protection

Banks are protected using various types of revetments and resistant lining materials. **Revetments** provide armouring, in the form of loose rocks and boulders, or container systems, such as wire baskets filled with rock (gabions). Banks can also be lined with concrete, asphalt, paving slabs or, where the engineered cross-section is rectangular, using vertical sheet steel piling. Plate 10.1 shows a heavily engineered urban channel which has been lined with concrete to prevent erosion. Lining a channel with concrete can lower the resistance to flow, possibly leading to problems of scour further downstream. For larger rivers, banks can be protected by laying down mattresses of concrete slabs connected by tough steel cables. This method has been extensively used along the lower Mississippi river, United States.

Spur dykes or **groynes** can be used to protect banks by deflecting flow away from vulnerable zones. They are built at an angle from the bank and are constructed from various materials, including stone, boulders, earth, gabions or pile. Thousands of stone spur dykes have been constructed along the lower Mississippi (Plate 10.2).

Bed protection

Armouring is often used to protect the channel bed from erosion. In-channel **grade control structures** can also be installed. These are of two basic types: sills and weirs. **Sills** are low, submerged structures, which are built at right angles to the direction of flow. They provide local fixed points that control the channel bed slope and water surface elevation to prevent degradation and headcutting. **Weirs** act as hydraulic controls, dissipating excess energy and reducing the energy slope

Plate 10.1 A steep, concrete-lined channel with grade control structures (weirs), Almaty, Kazakhstan.

(Plate 10.1). Protecting the bed from erosion reduces the available load and may lead to scour further downstream by sediment-hungry water.

River training

River training techniques have been used since as long ago as the late sixteenth century on the Yellow River in China (Przedwojski *et al.*, 1995). In Europe, aggrading glacially fed braided channels were among the first to be 'corrected'. One of the earliest and most successful examples was the work carried out on the Alpine Rhine (Switzerland) in the early nineteenth century. Before

then, the active braided channel had occupied a width of several kilometres. High rates of channel migration combined with frequent flooding meant that the floodplain could not be fully utilised. There was also a high incidence of waterborne disease, including malaria.

The main idea behind the works was to 'train' the river to flow in a deeper channel, reducing the incidence of flooding. This was achieved by confining the flow to a straight, single channel, using embankments and groynes to encourage deposition at the channel edges and to stabilise the channel in one position. Flow was concentrated along the centre of the channel, leading to

Plate 10.2 Spur dykes on the lower Mississippi, near Rosedale, Mississippi, United States. These encourage deposition, narrowing the channel and increasing flow velocity. In this way the depth of the navigation channel can be maintained. Photograph by Airphoto, Jim Wark.

deepening and an increase in channel capacity. This allowed floodwater and sediment to be rapidly transported downstream. By 1845, 12.5 million ha of the floodplain marsh had been drained, allowing increased rates of agricultural production (Downs and Gregory, 2004). Today the area is intensively cultivated.

Extensive training works have been carried out on many other rivers, including the Rhone, Danube and Mississippi. Spur dykes were constructed along the lower Mississippi to encourage deposition at the edge of the channel (Plate 10.2). This concentrates erosion of the channel bed, allowing the depth of the shipping channel to be maintained. In order to protect the opposite bank from erosion, it was necessary to install bank protection in the form of extensive concrete mattresses. Bed degradation is controlled using fixed weirs and sills (see above). Structures can also be installed to alter flow patterns on a more localised scale.

Dam construction

Dams are constructed for power generation, flood control, and for supplying water to irrigation schemes and urban centres. Today, very few large rivers remain unregulated, and the global volume of water stored in reservoirs now exceeds the volume of flow along rivers (Brierley and Fryirs, 2005). The scale of dams varies, from relatively small structures on tributaries to large dams that exceed 15 m in height. Gregory (1995) estimates that over 200 large dams are completed each year. However, there has been a decline in the rate of dam building in the industrialised world because many potential dam sites have now been developed.

Flow regulation dramatically alters the flow and sediment regimes. Flood peaks are reduced in magnitude and most of the sediment load is trapped in the reservoir behind the dam. Downstream from hydroelectric power stations there are often rapid fluctuations in discharge. Inter-basin transfers involve the movement of water across drainage divides, with the result that there is a net gain to some river systems and a net loss from others.

Emplacement of locks and weirs

The Rhine, Mississippi and Arkansas rivers have all been canalised to ensure a minimum depth of water for shipping and to increase the navigable channel further upstream. Canalisation involves installing dams across the channel to create a series of slackwater pools. Sluice gates, weirs and other control structures regulate the flow of water and vessels pass up and downstream through locks or, occasionally, ship lifts.

ECOLOGICAL REQUIREMENTS

In common with the geomorphological system, ecological systems are characterised by flows of energy and materials. They are also highly dynamic, being strongly conditioned by the hydrological and geomorphological environment and biological processes such as predation, competition, dispersal, migration, colonisation, ecological succession and extinction.

Biological communities

The biological communities that live in rivers, riparian zones and floodplains include many different types of plants and animals. Microbial organisms are important in primary production – the growth of plant material as chemical energy is stored through the process of photosynthesis – and in the breakdown of organic material and recycling of nutrients. The 'slime' that forms on submerged boulders and pebbles in river channels is a complex mixture of algae, bacteria, fungi and fine particles of organic and inorganic material (Closs *et al.*, 2004). Many plant species are anchored by their roots to the substrate, although floating plants can be found in backwater areas. Macroinvertebrates include insect larvae and nymphs, crustaceans such as freshwater shrimps, bivalve molluscs, gastropods (e.g. snails) and worms, while fish are the most notable vertebrates. Fauna that are associated with both terrestrial and freshwater environments include amphibians, reptiles and waterfowl, together with mammals such as otters, beavers or platypuses (depending on geographical location).

Habitats

Different biological communities are found in different **habitats**, the habitat being defined by the physical conditions. In-stream habitats are defined by channel morphology, bed and bank sediment, flow regime, sediment transport rates, in-channel and riparian vegetation, and woody debris. Many of these conditions fluctuate over short periods of time so it is the *range* of conditions that prevails, rather than an average condition that is important. For example, the physiology of many species of fish and invertebrates means that they cannot tolerate extremes of temperature. In order for a particular species to thrive in a given habitat there must be a tolerable range of conditions, a supply of food and an environment that can support all stages of its life cycle. Important wetland habitats are maintained by overbank flows and interactions between groundwater and surface water.

Within a reach of channel there are many small-scale variations associated with bedforms, scour pools, gravel patches, bars and shelter behind tree roots or large boulders. Associated with these differences are variations in flow depth and velocity. This creates a mosaic of **mesohabitats**, each characterised by different assemblages of species. Many species move between the different mesohabitats to feed, reproduce, find shelter during high flows, or hide from predators. Species diversity and abundance are usually higher where a range of different mesohabitats can be found.

Downstream changes

Along the length of a 'typical' river, recognisable changes can be seen in the nature and structure of biological communities as physical conditions change, from shallow, turbulent upland streams to deeper lowland channels. In the headwaters, variations in depth and velocity are more extreme and only those species that are able to withstand large variations in velocity and depth are able to survive. These make use of physical adaptations, such as a means of attachment (plants and macroinvertebrates).

Behavioural strategies include sheltering behind boulders or in pools (fish), or moving into the spaces in a gravel substrate (macroinvertebrates). Food resources are limited, because unstable substrates mean that few plants are able to establish themselves. The main source of food is provided by plant remains and the corpses and faeces of animals falling into the channel from adjacent slopes and riparian vegetation. Further downstream, plants are more able to establish themselves (although in deep channels with high turbidity this may only be at the edges of the channel) and biological communities become more diverse and abundant.

The fluvial hydrosystems concept

The fluvial hydrosystems concept (Petts and Amoros, 1996) recognises the important transfers of energy, materials and biota that occur within fluvial systems. These include longitudinal transfers along the channel, lateral transfers between the channel and floodplain, and vertical transfers between the channel and underlying alluvial aquifer (Figure 10.1).

- *Longitudinal transfers* are vital to many species that need to be able to move through the channel network. Migratory fish, such as salmon, swim downstream in their juvenile stage and live most of their lives in the ocean, before making the long journey back upstream to their spawning grounds. There are also downstream transfers of energy and biological material from the catchment area and through the channel network. These allow more productive and diverse biological communities to be sustained further downstream.

- *Lateral transfers* occur during overbank flows. Channels and floodplains become connected by floodwaters, enabling the lateral movements of water, sediment and biota that are vital to many species. In many fish species the timing of reproduction coincides with the timing of floods, when

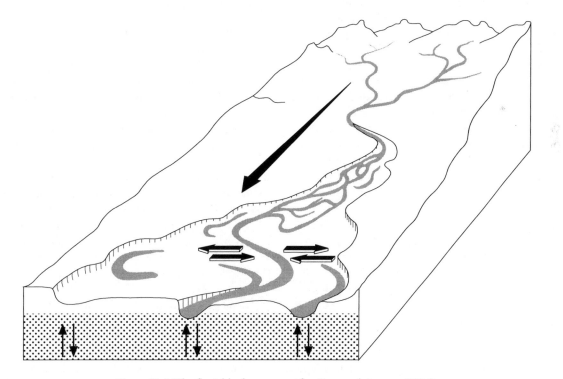

Figure 10.1 The fluvial hydrosystem. After Petts and Amoros (1996).

fish move onto the floodplain. Here they enter the sheltered environment provided by flooded hollows and abandoned channels, where spawning takes place. These environments also provide a refuge for many other aquatic species during flood flows. Periodic inundation of the floodplain creates unique wetland habitats that support rich biological communities.

- *Vertical transfers* between the alluvial aquifer and the overlying channel and floodplain are also important. This subsurface zone is important in supporting various groundwater fauna. Abandoned channels, which are isolated from the main channel at the surface, may still be connected periodically to the alluvial aquifer below the surface, allowing vertical transfers to take place.

ENVIRONMENTAL DEGRADATION

Human activity has led to the environmental degradation of numerous river systems. The fragmentation of river systems by dams, weirs and other structures seriously disrupts the natural functioning of physical and ecological processes. Together with declining water quality, this leads to a dramatic reduction in species diversity. The effects of human activity result in direct and indirect impacts on river systems. **Indirect impacts** are brought about by changes within the drainage basin that affect the flow and sediment regimes. Examples include deforestation, changing agricultural practices, urbanisation, mining and road construction. **Direct impacts** result from dam construction and channelisation, where modifications are made to the channel itself.

Basin-scale impacts

The basin-scale impacts of human activity on fluvial systems have been considered in Chapters 4, 5 and 9, which referred to deforestation, agriculture, mining and urbanisation. Deforestation, associated with agricultural development, has affected river systems for thousands of years. However, there has been an acceleration in the development of land and water resources over the last 100 to 500 years. Across much of Western Europe the intensity of agricultural production increased

dramatically after the Second World War. This intensification has included changes from grazing to arable land, the clearance and cultivation of riparian zones and increases in stocking densities. This has often increased the supply of fine sediment to river channels, leading to problems of siltation. Agrochemicals such as pesticides, herbicides and fertilisers have all increased pollution from agricultural runoff. Municipal sewage, industrial effluents and urban runoff also contribute to water pollution.

With increasing urbanisation and the move to a more industrialised society, the floodplains of major rivers in Europe and elsewhere are changing from agricultural to urban land use. This affects the flood hydrograph, increasing peak flows and necessitating further flood protection works. At the same time, because a greater proportion of incoming rainfall is rapidly diverted to rivers via drains, gutters and sewers, there is a reduction in groundwater recharge rates.

Impacts of dams

Changes to the flow regime

One of the most profound influences is the alteration of the flow regime by dams and other flow control structures. The life histories of many species have evolved in response to natural flow regimes. As you have seen in previous chapters, flow affects all aspects of the physical habitat, including the shape and size of the channel, the spacing of riffle and pool habitats and nature of the channel substrate. Even at small scales, variations in velocity and shear stress across the channel bed affect the distribution of plants and macro-invertebrates within the channel. For most fish species, the timing of life events such as reproduction and spawning can be linked to the flow regime. The timing of rising flows are important for fish that move out onto the floodplain to spawn. Other triggers include day length and temperature. Under an altered flow regime, these may no longer be synchronised with natural variations in flow.

The impacts of flow regulation on a given river system are dependent on several factors, including the number and size of dams, the distance downstream from an impoundment and the proportion of the upstream drainage basin area that is regulated. Further downstream, the effects of flow regulation may be reduced to some extent by unregulated tributaries joining the main channel. The type of operation, for example hydroelectric power generation, is also significant. One of the main impacts is the reduction in flood peaks, which play a vital role in the life cycle of many species. Irrigation schemes can even result in a reversal of the flow regime, when seasonal flood flows are impounded and later released to water crops during dry months.

Downstream from hydroelectric plants the flow can vary considerably in just a few hours, as electricity is generated to meet daily fluctuations in energy demand. Such releases can have a serious impact on the temperature regime. During relatively warm conditions, water released from the base of a dam originates from the cooler depths of the reservoir. This can result in severe thermal shocks to fish and other species downstream from the dam. In July 1976 high mortalities occurred among grayling (a fish species) in the River Ain, a tributary of the Rhône, because of twice-daily releases of cold water from upstream reservoirs (Bravard and Petts, 1996). Seasonally modified temperature regimes also affect life cycle patterns and coldwater releases have been found to delay spawning by up to thirty days in some fish species (Zhong and Power, 1996).

In the Western world, **inter-basin transfers** are becoming more common as most potential dam sites have now been developed. This involves diverting water from one drainage basin to another, via a pipeline or canal, to enhance flow in the receiving drainage basin for water supply or irrigation. Downstream from the abstraction point, a '**compensation flow**' is released into the channel to meet the needs of downstream users, and to provide sufficient water to dilute sewage and industrial effluents that are discharged into the channel. However, this is usually much less than the natural flow and may not be sufficient to meet environmental needs. In dryland environments, the receiving river can be transformed from an ephemeral to a perennial channel. This is the case with the Great Fish River in South Africa, which has changed from a series of unconnected pools in the dry season to a perennial river as a result of the Orange–Fish inter-basin transfer. Such transfers have major implications for existing plant and animal communities that have evolved to a highly variable flow regime. In fact, non-native 'exotic' species may thrive under the new flow regime. Regulation of flows in some Australian rivers is thought to favour non-native carp and mosquitofish (Bunn and Arthington, 2002). Despite precautions, exotic organisms, including parasites, bacteria and fish, are often unintentionally transferred between basins via connecting pipelines. This is how Orange River fish have made their way into the Great Fish River in South Africa.

Reduced connectivity

Longitudinal connectivity is greatly disrupted by dams, locks and weirs. In many cases migratory pathways are blocked, even by relatively small structures. Migratory species including shad, lamprey and eels have disappeared from the Rhône in France (Bunn and Arthington, 2002). Fish ladders are often installed at the site of dams to allow fish to bypass the dam by swimming up through a flight of pools. For various reasons these structures are not always successful, meaning that reduced numbers of fish are able to make the journey upstream.

Lateral connectivity is reduced by the dramatic reduction in the frequency, extent and duration of overbank events. Bravard and Petts (1996) cite the case of the Volga where the duration of floodplain inundation has decreased, from fifty to seventy, to ten to fifteen days a year, since flow regulation began. This has had a major impact on fish populations, since a minimum of forty days is needed for the growth and development of juvenile fish before their descent to the Caspian Sea. A reduction in flood inundation also threatens wetland areas, particularly where land drainage is carried out. The Macquarie Marshes in eastern Australia, a wetland reserve for birds, has been reduced to between 40–50 per cent of its original size by flow diversions and weirs (Kingsford and Thomas, 1995). Lateral connectivity is further reduced by levees and embankments and artificially deepened channels (see below). Vertical transfers are also affected, since recharge of the underlying alluvial aquifer takes place when floodwaters inundate the floodplain. Reduced rates of aquifer recharge lead to a drop in groundwater levels, further contributing to the drying out of wetland areas.

Impacts of channelisation

Instability problems

Channelisation programmes have significantly modified tens of thousands of kilometres of river channels (Brookes, 1985). These modifications often lead to instability within the engineered reach and in the reaches upstream and downstream from it. Changes to the channel slope, width, depth or roughness all affect channel hydraulics. Feedback mechanisms lead to adjustments as the channel tries to find a new equilibrium. For example, in a channel that has been enlarged for flood control, there will be a reduction in velocity and unit stream power at low flows. This will result in deposition along the reach, meaning that the channel has to be re-dredged on a regular basis to maintain its capacity.

Modifications to the channel slope, made by creating artificial cutoffs, gravel mining or dredging, can have the most dramatic effects because of the resultant increase in stream power. Incision often occurs upstream from the artificially steepened reach. Erosion is concentrated at the break in slope between the gentler upstream reach and steeper engineered reach. Upstream incision then takes place as a series of headcuts migrate upstream, although the rate of incision decreases in an upstream direction. In severe cases, banks become unstable, resulting in collapse and channel widening. Bridges and other structures can also be undermined. The additional sediment that is produced causes further problems of aggradation downstream.

Major instability problems resulted from the removal of sixteen meander bends along the lower Mississippi between 1929 and 1942. It shortened the channel by 220 km and led to excessive channel erosion, necessitating further intervention. Prior to the installation of bank protection on a massive scale, erosion was removing 900,000 m³ of bank material a year (Bravard and Petts, 1996). Upstream degradation has also led to extensive deposition in the engineered reach, in the form of bars.

Geomorphological response times vary, being dependent on the type of work carried out, and the extent to which unit stream power, sediment supply and vegetation cover are affected. In some cases it may take up to 1,000 years for the channel to reach a new equilibrium form (Brierley and Fryirs, 2005).

Ecological impacts

The abundance and diversity of different species tends to be greatly reduced in engineered channels as a result of limited connectivity and habitat availability (Figure 10.2). Dredging and snagging remove geomorphic structures such as riffles, pools and bars, and disturb the structure of bed sediment. The uniformity of engineered channels provides little variety, affecting the viability of certain species. This is particularly true of concrete-lined channels, which have very little ecological value.

During high flows, stream velocities may be higher than some species can withstand. Deepened channels and levees increase channel capacity, greatly concentrating high flows. Opportunities for shelter are reduced within the channel, and levees prevent access to calmer waters on the floodplain.

Water temperatures can increase to intolerable levels during low flows. Enlarged channels may not provide a sufficient depth of flow, a problem that is exacerbated by the removal of pools and the shading effects of riparian vegetation.

Aesthetic impacts

As well as affecting the morphology and behaviour of river channels, channel engineering works often have a negative impact on the appearance and amenity value of the channel. Part of what makes natural rivers pleasing to look at is the amount of variety one can observe, even over a short distance. Variations in depth, velocity, slope and sediment size, associated with forms like riffles and pools, bends, bars, rapids, trees and other vegetation, all combine to create an interesting environment. By contrast, engineered reaches can be very monotonous in appearance, with their straight channels and uniform cross-section, cut off from the adjacent floodplain by embankments and levees.

A heavily engineered channel like the one shown in Plate 10.3 resembles little more than an open drain. This particular example is to be found hidden beneath the multiple flyovers of Birmingham's famous Spaghetti Junction in England. However, many more examples can seen in parks and residential areas. They do little to enhance the urban environment and have no amenity value.

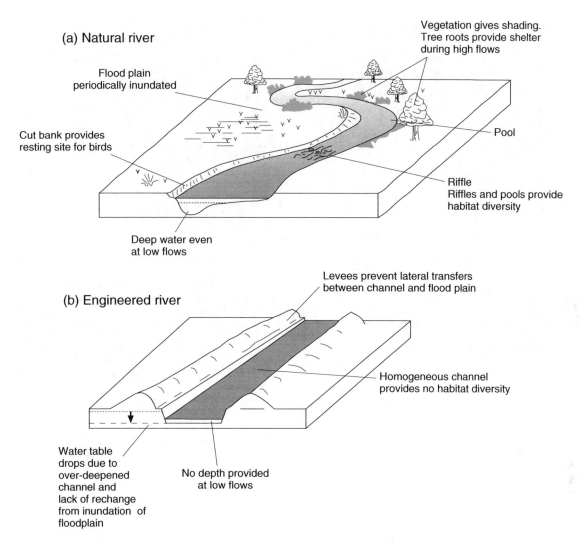

Figure 10.2 The habitat diversity provided by (a) natural and (b) engineered channels.

ENVIRONMENTAL APPROACHES TO CHANNEL ENGINEERING

Changing management priorities

Environmental awareness increased substantially over the last part of the twentieth century. This was brought about by increased understanding of the environment, together with concern for loss of habitat, loss of species diversity and the implications of ecosystem decline for human health. River systems play a vital role in sustaining human health, and their degradation also has socio-economic implications for the value of river systems in commercial and amenity terms. The 'health' of a river can be described in terms of the quality of its habitat structure, hydrological characteristics, ecological function and water quality. Growing concern for river health is reflected in statutory requirements, where the preservation of

Plate 10.3 The heavily engineered River Tame as it flows beneath Spaghetti Junction, Birmingham, England.

ecosystems is a priority, together with human requirements. Central to this is a holistic approach where, at whatever scale is being considered, the functioning of the whole drainage basin is taken into consideration.

Basin management involves prioritising and dealing with many different, and often conflicting, demands. These include meeting ecological requirements; managing channel instability; providing a supply of water for agriculture, urban centres, industry and power generation; flood management; navigation; recreation; and the needs of local communities.

Appropriate management strategies are also dependent on the condition of the river itself, in terms of both river health and channel dynamics. On a global scale, very few rivers are in a pristine or near-pristine state. Even where channelisation has not been carried out, land use changes within the drainage basin have often resulted in altered flow and sediment regimes, as well as a decline in water quality. The first priority should be in the **preservation** of channel reaches that are in a pristine or near-pristine condition. Second, in modified reaches where there is potential to return the river to a more natural state, various channel **restoration** strategies can be implemented. The lowest priority are cases of severe environmental degradation, where the only option may be to leave the channel in a state of **dereliction** and focus resources elsewhere.

Information requirements

In order to integrate the many different priorities and work with, rather than against, the natural functioning of river channels, it is essential to start by understanding the condition that they are in. Environmental assessment is a very important aspect of channel management and many countries have legislation that requires pre-project assessment to be carried out.

A range of environmental assessment techniques have been developed for different purposes and for varying levels of detail. There are usually three levels of pre-project environmental assessment (Downs and Gregory, 2004). These are briefly outlined below.

Level 1. Basin-scale assessment of river channel condition

The first stage involves compiling basin-scale inventories of channel morphology and habitat condition along the length of the channel. This allows identification of reaches of high conservation value, the location of past channelisation works, areas of instability, the morphology of different reaches and the habitats associated with them. Assessment usually includes a stream reconnaissance survey and a baseline habitat survey.

The **stream reconnaissance survey** involves combining basic measurements of channel geometry with qualitative assessments of channel conditions, including valley characteristics, land use, connectivity, and in-stream and riparian vegetation. Standard formats have been devised for collecting this information, such as the *Stream Reconnaissance Handbook* (Thorne 1998), although skilled interpretation is required in the field. Field observations are supplemented with information from maps and aerial photographs.

Baseline habitat surveys provide further information on the 'health' of the river in terms of its water quality, geomorphic structure and connectivity. These are based on the hierarchical links that exist between channel processes, form, habitat and associated biota.

The character of different in-stream habitats is determined by interactions between channel morphology; bed and bank sediments; the flow regime; flow hydraulics (mainly by the combinations of velocity and depth conditions); sediment transport characteristics; and riparian vegetation (Downs and Gregory, 2004). Certain groupings of biota are associated with different in-stream habitats. Surveys can be 'top down', starting from morphological surveys, or 'bottom up', where data from biotic surveys are grouped (Downs and Gregory, 2004). Top-down approaches are based on the concept of **hydraulic** or **physical biotopes** (Rowntree and Wadeson, 1996; Newson *et al.*, 1998). These are habitat 'units' that are defined by combining the physical features of the channel with flow character. On this basis, hydrogeomorphological units such as 'riffle', 'rapid' and 'pool' can be distinguished. In this way, links can be made between habitats, flow types and species requirements (Newson *et al.*, 1998).

Level 2. Basin historical analysis

River channels are dynamic, and it is vital to assess each reach in terms of the controls on channel adjustment, sediment dynamics (supply, transport and storage), whether the reach is stable or unstable and its potential sensitivity to change.

In order to understand the condition of a given channel reach, it is essential to link a local site, such as a channel reach, with the wider context of the drainage basin and its history (Gilvear, 1999).

Basin history can be reconstructed using flow records, aerial photographs, maps, historical records, floodplain stratigraphy and vegetation (Chapter 9). Interpretation of this information provides insight into the ways in which changes in the flow and sediment regimes reflect changing environmental conditions within the basin. Through this understanding it is possible to determine the sensitivity of a given reach to change and to develop appropriate management strategies.

Using a **fluvial audit**, the condition of the reach – channel stability, sediment movement and morphological stability – can be related to sediment dynamics in the catchment (Sear *et al.*, 1995). This is constructed using information on sediment sources, pathways and characteristics.

Level 3. Reach-scale analysis of the sensitivity of channel hazards and assets

In order to determine the risk associated with channel-related hazards, such as flooding or lateral erosion, detailed reach-scale analyses are carried out. This requires considerable expertise and intensive fieldwork, so is feasible only for a small number of reaches. These are carefully selected on the basis of the preceding catchment and historical assessments, and should be representative of similar reaches. Examples of the field measurements carried out include planform mapping, surveys of bed topography, measurement of velocity fields, sediment transport rates, rates of lateral erosion, and analysis of the structure, composition and dominant failure mechanisms of the banks.

Reducing the impacts of channel engineering works

Channel engineering can now be carried out using more environmentally sensitive techniques. These reduce the impact on the natural function and morphology of the channel, and promote greater ecological diversity.

Distant flood banks and multi-stage channels

Where space allows, a more environmentally sensitive approach is to use distant flood embankments, which are set back from the channel (Figure 10.3a). These allow the river to inundate its floodplain but provide flood protection at the same time as maintaining important channel–floodplain interactions. Distant embankments do not need to be as high as those built adjacent to the channel to provide the same level of protection and the channel itself remains undisturbed.

Two-stage channels are designed to accommodate a range of flows, with the low-flow channel being preserved within an excavated flood channel (Figure 10.3b). The berms allow enhanced interactions between surface water and groundwater because the frequency of inundation is increased, and also because the water table is closer to the lowered surface. Problems can arise if vegetation growth on the berms remains unchecked, as this will increase the roughness of the flood channel, reducing velocity and channel capacity. Another problem can be the build up of silt on the berms.

Partial dredging

Rather than dredging along the entire length of the engineered reach, partial dredging reduces the impacts on in-stream ecology. This involves dredging only the central part of the cross-section at intervals along the channel. For example, the depth could be increased at shallower riffles, leaving parts of the bed undisturbed. Careful design is required to ensure that instability problems do not result. Partially dredged sections of channel can rapidly become infilled with sediment, necessitating regular maintenance.

Revetments using 'soft' engineering techniques

A wide range of bank protection measures can now be used. These range from 'hard' engineering structures such as rip-rap and concrete to 'softer' or 'biotechnical' engineering solutions. For situations where bank erosion is a serious threat – in urban areas and where channels run close to roads or railways – hard engineering structures are most appropriate. However, in less critical settings, soft engineering provides a more environmentally sensitive alternative.

In high-energy environments the placement of a stone 'apron' protects the toe of the bank from erosion. The spaces between the interlocking stones allow plants to become established and provide habitats for in-stream fauna. Where the specific stream power is lower, natural plant materials can be used to protect banks by placing a line of stakes along the bank and interweaving them with branches (Figure 10.3c). Fast-growing species such as willow may be used as stakes to create a living barrier. Geotextiles include materials such as jute, coconut fibre and brush mattresses (woven from brushwood). These provide bank protection whilst stabilizing vegetation becomes established. Reeds and grasses also provide some degree of protection and encourage deposition of fine sediment at the channel margins. Although soft engineering is less expensive than building traditional hard engineering structures, it can be labour-intensive to install, takes time to establish, and requires regular monitoring and maintenance. Biotechnical revetments are less durable than more traditional approaches, meaning that these approaches are not suitable for higher-energy environments.

Controlled flooding of wetland areas and flood basins

An alternative method of flood control is to divert floodwaters into temporary storage on floodplains, wetland areas or artificial flood basins, in order to reduce the size of the flood peak further downstream.

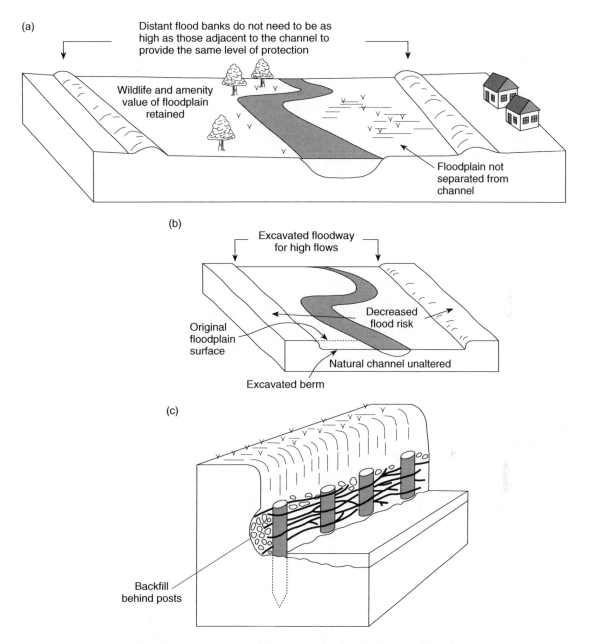

Figure 10.3 Environmental management techniques. (a) Distant flood embankments. (b) Multi-stage channel. (c) Bank protection using woven fencing. (c) adapted from Lewis and Williams (1984).

Wetland areas can be reinstated by changing floodplain land use to grazing or other low-risk activities. Control structures such as sluice gates may be used to divert floodwater onto the floodplain to reduce the flood risk to urban centres and valuable agricultural land further downstream. There are also many examples of flood basins, which are constructed to provide a temporary store for floodwater. Controlled by sluice gates, water is then gradually released as the flood recedes.

RIVER RESTORATION

River restoration is defined as the restoration of a river to a former or original condition (Downs and Gregory, 2004: 240). This is where techniques are applied to assist recovery and accelerate the re-establishment of natural physical and ecological processes. It is often not possible to restore a river fully to its former condition, owing to changes in the drainage basin, which alter the flow and sediment regimes. However, a new condition can be established in which natural function is restored.

Restoration techniques can be passive or active. **Passive restoration** involves addressing factors that are preventing recovery, such as activities within the basin that adversely affect water quality, sediment and flow regimes. Examples of passive restoration include the establishment of buffer strips – zones of natural floodplain vegetation alongside the channel – and reinstatement of the flow regime. **Active restoration** is when specific modifications are made to accelerate recovery, such as the morphological reconstruction of meanders, riffles and pools.

Buffer strips and riparian zone management

Natural riparian vegetation fulfils a number of important roles, including bank stabilisation, nutrient regulation, filtering of sediments, shading, a source of large woody debris and a nesting site for birds (Gordon et al., 2004). However, intensification of agricultural practices has resulted in the widespread removal of riparian vegetation to provide access to water for grazing animals, and to increase the cultivated area. Livestock can have a significant influence on river channels when banks are trampled and overgrazed. This leads to instability and channel widening, with a resultant increase in the volume of sediment entering the channel. It is estimated that livestock grazing has altered approximately 80 per cent of stream and riparian ecosystems in the western United States (Belsky et al., 1999). By fencing off riparian areas, livestock access can be restricted to allow natural vegetation to re-establish.

Buffer strips are strips of channel-adjacent land that run parallel to the river. Natural vegetation is encouraged to re-establish in these zones, which, together with the channel, provide corridors across the landscape. Buffer strips have been shown to filter fine sediments, which may carry contaminants (Chapter 5), and allow dissolved chemicals to be taken up by plants (Large and Petts, 1994). Flood peaks are reduced when floodwaters are stored on the floodplain, and there are also significant benefits for groundwater quality due to increased recharge. The effective width of buffer strips varies from channel to channel and is partly dependent on the size of the channel, while wider areas are needed to trap fine sediments such as silt. Typical widths range from 15 m to 80 m (Large and Petts, 1994).

In South Africa, exotic riparian vegetation, such as the Australian Eucalyptus, creates major problems because of the vast quantities of water it consumes. Since the Working for Water programme began in 1995 over a million hectares of thirsty invasive alien plants have been cleared. If these plants were not removed, 30 per cent of runoff would be lost to rivers within ten to twenty years, and 74 per cent within twenty to forty years (DWAF, 2001). The programme has been highly successful in restoring flow to rivers and in creating large numbers of jobs.

Restoring the flow regime

Flow regulation is widely recognised as being one of the main causes of environmental degradation in rivers and is a priority issue in river management worldwide. Flow restoration involves adjusting regulated flow regimes and returning diverted flows to rivers. This can be controversial, particularly in water-stressed regions, where the available water supply is barely adequate to meet competing demands for human needs.

Determining what is an 'ecologically acceptable' flow regime is a difficult task. In addition to determining a minimum acceptable flow to sustain biological communities, flow variability is necessary to maintain ecosystem structure and function. Flood pulses are a particularly important aspect of natural flow regimes. Flood flows with higher return periods are needed to rejuvenate the floodplain system, while more frequent, channel forming floods maintain the morphological features of the channel (Petts and Maddock, 1996). As a general rule, this corresponds to a flow with a one-and-a-half-year return period for stable alluvial channels, although there are wide variations (see Chapter 3, pp. 32–33). Considerations such as the dam outlet design impose limitations on the magnitude of artificial floods.

To maintain salmon fisheries in regulated rivers, **pulse releases** are made from dams to stimulate migration and to flush out fine sediment from spawning gravels. The timing of such flows has to be carefully planned to fit in with life cycles. Flow regulation can also be used to enhance natural floods. A natural one-in-five-year event on the Murray River in Australia, which occurred in 2000, was enhanced by three releases from a headwater dam. This increased the duration of the flood peak and extended the recession curve, maintaining suitable conditions for breeding birds in the Barmah–Millewa Forest. As a result there was an increase in the number of species and bird numbers (Gordon *et al.*, 2004).

An interesting experiment was carried out on the Colorado River, United States, in April 1996, when a controlled flood flow was released into the Grand Canyon, from the Glen Canyon Dam, over a seven-day period. The aims of the experiment included remobilising sediment that had built up where unregulated tributaries joined the main channel, and building up sandy beaches. The controlled flood provided research opportunities for scientists from a range of fields and was closely observed. Further information can be found in Box 10.1.

Dam removal may be a viable option, and many smaller structures have been decommissioned in the United States and elsewhere. A major problem associated with dam removal is the large volume of sediment that is trapped behind the dam, since releasing it into the channel would have serious consequences downstream.

Morphological restoration

Passive restoration – environmentally sensitive flow regimes and buffer zones – enables the full or partial recovery of formerly degraded channel reaches and river systems. 'Recovery' rarely involves a return to the original state, since land use and other changes in the catchment are likely to have altered the flow and sediment regimes. However, the morphology and function of the channel are restored, providing the diversity needed to sustain river health.

In the case of lowland channels with limited stream power, slow rates of adjustment mean that natural recovery could take several centuries (Gordon *et al.*, 2004). The process of recovery can be accelerated through **structural restoration**, the main aim of which is to restore structural diversity to a given reach. Morphological reconstruction is also used on upland reaches, where previous channel modifications have led to instability. Structural restoration is expensive and labour intensive, meaning that it is feasible in only a limited number of cases. There are also adverse impacts for in-stream and riparian biota, which will experience a certain amount of disturbance during the construction phase. A number of different techniques are employed, but all require a good understanding of fluvial dynamics, and morphological structure and function. Careful design is necessary to ensure that instability does not result, and that the newly modified channel is self-maintaining. Particular attention needs to be paid to flow resistance and rates of sediment transport.

In-stream structures

In-stream structures are installed to recreate physical diversity in channelised reaches. These structures create areas of flow divergence and convergence, modifying conditions at the bed and encouraging localised erosion and deposition. Prior to installation, a detailed assessment of the reach is essential, in order to understand the causes of channel degradation and the needs of fish and in-stream biota. Care is required in installing these structures as they can work *against* the river if incorrectly applied (Downs and Gregory, 2004). For example, in higher-energy settings, it is important to ensure

Box 10.1

CASE STUDY: EXPERIMENTAL FLOODING IN THE GRAND CANYON

Since the 1970s, the flow in the Grand Canyon has been regulated by the Glen Canyon Dam. Before the dam was built, the natural flow regime was characterised by summer floods which built up sandbars and sandy beaches, removed the build-up of coarse material from debris flows and periodically stripped vegetation from the banks. Since completion, the dam has been operated for hydroelectric generation. This has greatly altered the flow regime with daily, rather than seasonal, fluctuations in flow and a substantial reduction in flood peaks. Because so much fine sediment was trapped behind the dam, sandbars and beaches began to disappear. At the same time, coarse material from side canyon debris flows and landslides built up in the channel, because the regulated flow was no longer sufficient to transport it.

In 1983 an emergency release of 2,750 m^3s^{-1} had to be made because rapid snowmelt was filling the reservoir to dangerously high levels. The results of this flood were quite surprising to scientists and river guides who were familiar with the river. Many of the disappearing beaches were now replenished with fine sand, and some riverine habitats were improved. The next three years were wetter than average and subsequent, smaller, releases started to erode the beaches again.

Although sand was provided by two unregulated tributaries that joined the Colorado downstream from the dam, much of it was building up along the bed of the channel and the beaches were not being replenished. One of the main aims of the 1996 release was to remobilise sand from the bed and redeposit it on the beaches. In addition, coarse sediment that had built up in the channel would be transferred downstream. A further aim was to flush out invasive alien vegetation, such as tamarisk, which had encroached on the channel.

March was chosen for the release because it was before fish spawning and after overwintering birds had left the canyon. Concerns had been raised about the potential impacts of the flood on fish, but most fish appear to have moved to sheltered areas along the banks as the flood waters rose. There was little damage to the algae and invertebrates that provide an important food source. Over a seven-day period, water was released at a rate of 1,270 m^3s^{-1}. The flood deposited significant amounts of sediment above the normal high water line, building up many of the sandy beaches. It also widened the two largest rapids. However, there was little impact on the encroaching vegetation. Several other experiments have since been carried out on the basis of the findings from the 1996 flood release.

Source: The information provided here is based on Collier *et al.* (1997).

that erosion does not occur where flow is deflected towards the bank.

Weirs or low dams create an area of ponding immediately upstream from the structure. As the flow cascades over the weir, scouring of the bed occurs at its base (Figure 10.4a). Weirs are constructed using natural materials, such as wood or stone, and can also be angled to encourage localised areas of scour and deposition, as shown in the diagram. Obviously it is important to ensure that these structures do not impede the upstream migration of fish and other species. Such structures need

careful design and must be 'keyed in' to the bed or bank so that they are not undermined by erosion (Downs and Gregory, 2004). This could lead to subsequent failure of the structure.

Flow deflectors are widely used to promote localised scour and deposition, to deepen the channel and to divert flow away from vulnerable banks. Figure 10.4(b) shows just one example, where the flow accelerates through a narrow gap to increase bed shear stress and deepen the channel. At the same time, recirculating flow deposits material at the edges of the channel.

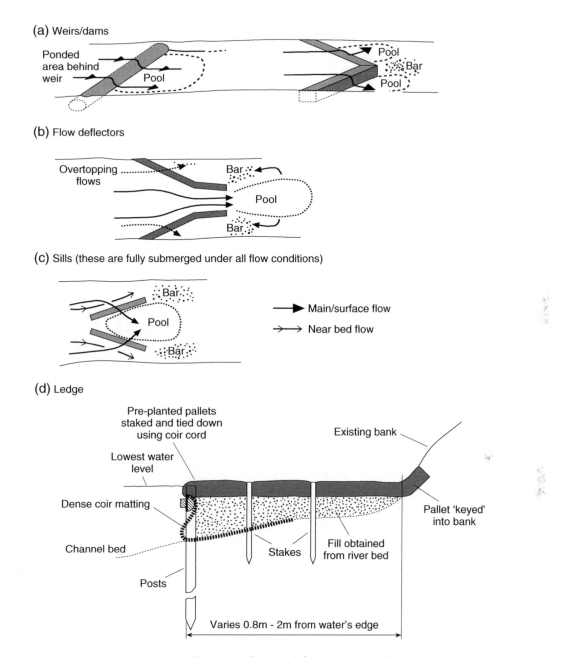

Figure 10.4 In-stream structures. (a) Weirs or low dams. (b) Flow deflectors. (c) Sills. (d) Aquatic ledge. (a), (b) and (c) adapted from Hey (1994), (d) adapted from RRC (1999).

Other configurations can be used to create different patterns of flow convergence and divergence, according to the specific needs of the application. A critical design consideration is setting the height of flow deflectors relative to the normal range of water levels. For example, if live willow logs are set too high, excessive growth could lead to flood obstruction (River Restoration Centre, 1999). Bank-attached **groynes** or **spur dikes** are often used for bank protection. These project at an angle from the bank and deflect the *thalweg* away from the eroding bank. Constructed from materials that include stone or woven fencing, these structures encourage deposition. Over time the colonisation of vegetation stabilises the deposits.

Sills are fully submerged, even during low flow conditions, and modify near-bed flows (Figure 10.4c). Depending on the characteristics of the specific reach, boulders can be placed to provide shelter for fish and anchorage for plants. Boulders also affect flow patterns, to create small-scale variations in the channel substrate and provide a means of dissipating excess energy. The significance of large woody debris is becoming increasingly recognised. Emplacement of bank-attached tree trunks and root boles greatly increases microhabitat availability and provides shelter for fish at high flows.

Modifying channel dimensions

Rivers that have been engineered for navigation or flood control have often been over-deepened and/or widened. Various techniques are used to reduce channel width. For example, in-stream structures, such as flow deflectors, can be installed to encourage deposition along one or both banks of the channel. Where rates of sediment supply are low, **aquatic ledges** or **berms** are installed (Figure 10.4d). As well as narrowing the channel, these provide a valuable wetland habitat during low flows.

The steep, canyon-like banks of over-deepened channels can be reprofiled mechanically, using a bulldozer or similar. Excavation may also be used to restore asymmetric variations in channel cross-section. Over-deepened reaches are sometimes partially infilled with gravel, care being taken to ensure that it is of a size that will not be flushed downstream. To restore channel–floodplain interactions, excavation of the floodplain surface on one or both sides of the channel creates a new, lower elevation floodplain adjacent to the channel.

With all these designs, it is also important to recreate the natural variability that occurs along channel boundaries, including overhangs, variations in bank angle and irregularities along the line of the bank.

Meander restoration

As well as providing a variety of flow conditions, meander reconstruction lengthens the channel and thus increases the area of channel habitat. Meanders also reduce channel gradient and, in turn, velocity and transport capacity. Re-meandering, if carefully designed, can therefore reduce the instability of artificially straightened reaches. Meanders have been reconstructed on a number of previously straightened lowland channel reaches in Germany, Denmark, the UK and elsewhere. Since 1994, restoration of over 20 km of the River Brede in South Jutland, Denmark, has re-created a meandering channel that flows across a 500 m wide floodplain. A smaller-scale example of meander restoration in an urban setting is the 1 km re-meandered reach of the River Skerne, England (Box 10.2).

Careful design of meander bends is essential if instability is to be avoided. In some cases, the former meandering course can be reconstructed using old maps, aerial photographs, or other evidence. For example, the original course of the River Cole in Oxfordshire, England, was traced from historic maps, from remnants of meanders on the floodplain and avenues of willows that lined its former course. Another method is to copy a new meandering course from one or more **reference reaches** elsewhere along the same, or a similar, river. A rough idea of meander dimensions can be obtained using established relationships between channel width and meander wavelength (Chapter 8, pp. 138–141). However, other variables such as slope, channel substrate and rates of sediment transport also need to be taken into account. After meander construction there may be some migration and adjustment of the channel, so bank protection may be necessary where erosion is undesirable.

Box 10.2

CASE STUDY: THE RIVER SKERNE RESTORATION PROJECT

The River Skerne in Darlington, north-east England, was once a naturally meandering channel. This was before the town expanded, when the river was deepened and straightened to provide flood protection. A number of industrial sites along the river had polluted it in the past, and much of the floodplain had been affected by the tipping of industrial waste. The restoration reach is located in the Darlington suburb of Houghton-le-Skerne, where a small section of floodplain had survived the urban encroachment and industrial tipping that had occurred elsewhere. Prior to restoration, this straightened, over-deepened reach was typical of many channels in urban environments and had little ecological, aesthetic or amenity value. Its main functions were to act as a drain for floodwater, providing a conduit for discharges from outfalls, land and road drains. Further downstream the river flowed through an artificial canyon, formed from industrial tippings that had been made right up to the channel margins.

Restoration of this reach was carried out in 1995–96 to demonstrate the techniques and benefits of lowland river restoration in an urban setting. The project was led by the River Restoration Centre (then the River Restoration Project) in partnership with government agencies and voluntary organisations. One of the main aims was to show that it was possible to create a channel with natural features at the same time as providing protection from flooding for nearby residential and industrial areas.

The urban setting presented a number of challenges. Although the restoration reach flows through an urban park, the main sewage and gas pipelines for Darlington run alongside the channel. In addition one of the buried electricity mains crosses the channel. Careful planning was therefore necessary to reduce erosion and provide protection in the vicinity of these utilities. Protection was also necessary to prevent the erosion of contaminated land, and a number of revetments were installed. These included willow spiling, where flexible willow poles were tightly woven around vertical stakes. A rock matrix was used to line the bank prior to the installation of fencing, and live willow was planted behind the fencing to provide stabilisation. Other techniques used crushed rock, logs, geotextiles and live willow – further details can be found in the River Restoration Centre's *Manual of Techniques*.

Thousands of tons of earth were excavated in creating the new river and lowering the floodplain. This was used to re-landscape the area to hide nearby industrial buildings. Along a 1 km reach, four new meanders were created on the south side of the old, straightened course (Figure 1). These were carefully profiled to increase the variety of different bank forms and to enable public access to the channel. Two new backwater areas were also created to increase the range of available habitats. Seasonal flooding now occurs within the area of the restored floodplain, although this is contained within the excavated area. Further upstream and downstream, where it was not possible to morphologically reconstruct the river, a number of habitat enhancements were made. These included the installation of a stone riffle, aquatic ledges and in-channel deflectors. In total, 2 km of the river were rehabilitated.

The concerns and needs of the local community were taken into consideration through public consultation throughout the project. One of the main concerns was accessibility, and a network of pathways were constructed as part of the project. The restoration of the Skerne has been enthusiastically received by the local community and provides improved management of floods, droughts and water quality. Within only two years of completion, pools and shoals had been formed by the in-stream structures. Previously absent or uncommon species, such as swans, fish, dragonflies and water voles (a protected species) have also been recorded. Along with other demonstration projects carried out on the River Cole in Oxfordshire and the River Brede in Denmark, a wealth of information has been provided. This has formed the basis for many subsequent restoration projects that have been carried out elsewhere.

(Continued)

Box 10.2

CASE STUDY: THE RIVER SKERNE RESTORATION PROJECT—CONT'D

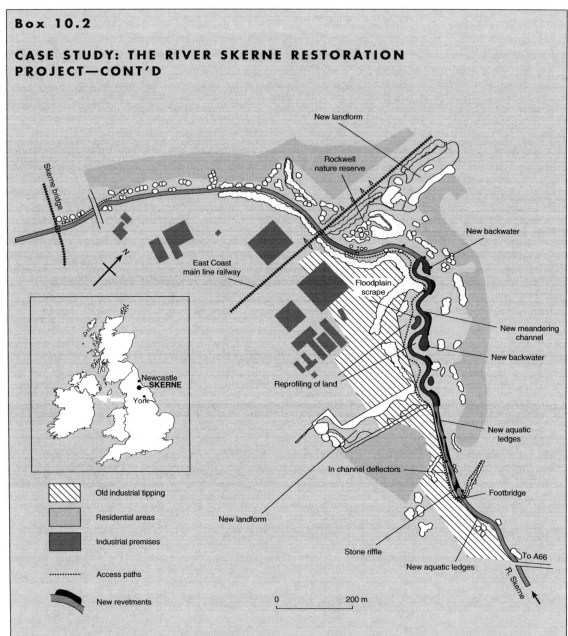

Figure 1 The restored reach of the river Skerne, Houghton-le-Skerne, Darlington, north-east England. Adapted from RRC (1999).

Recreation of pools and riffles

Riffles and pools provide a means of dissipating excess energy as well as increasing habitat diversity. Various methods are used, including the excavation of bed sediment to create pools, the placement of gravels or cobbles to form riffles or through the use of appropriately spaced in-stream structures. For any of these methods to be successful, an understanding of morphologic controls and flow-sediment interactions is essential to prevent newly created pools from silting up, riffles from being eroded, or the spaces between newly laid gravels becoming infilled with fine sediment.

Steps and pools

These structures help to oxygenate the water, but are only appropriate in steep reaches where steps and pools would naturally occur. They may also impede the upstream migration of fish, so may not always be suitable. Weirs are constructed from stone walls, gabions, wooden poles or tree trunks. Water ponds up behind the weir before cascading over the crest to scour another pool at the base of the weir. Bank protection is usually necessary to prevent weirs from being removed by erosion.

CHAPTER SUMMARY

The development of land and water resources has had an adverse effect on many rivers, modifying flow regimes, water quality, morphology, and physical and ecological function. Direct impacts on river channels result from dam construction and channelisation. Indirect impacts are brought about by activities within the drainage basin, such as agricultural development, urbanisation, and land drainage schemes. Traditional management has focused on human needs, with little regard for the physical, chemical and biological condition of river channels. River channels are increasingly fragmented by dams, weirs, levees and other structures. These block migratory paths and prevent important transfers between channels and floodplains. Altered flow regimes have many adverse effects, such as disrupting the timing of life-cycle events and reducing the extent and duration of flooding. Channelisation greatly reduces the diversity of habitats within river channels,

meaning that few species can survive. The links between ecosystem health and human health are increasingly recognised, and growing environmental concern has led to a shift in management priorities. New 'soft' engineering techniques have been developed to reduce the impact of channel interventions, at the same time as addressing human needs. There are now many examples of formerly degraded rivers that have been restored by applying techniques to assist recovery. Careful design is necessary to ensure that the structures installed are appropriate for the specific application. Channel instability problems can result if these techniques are incorrectly applied. Further research is also needed to improve on existing designs. It is rarely possible to return a river system to its original condition, but its natural function can potentially be restored. This can be achieved by restoring flow regimes, managing land use and reconstructing channel morphology.

FURTHER READING

Introductory texts

Mant, J. and Janes, M., 2006. Restoration of rivers and floodplains. In: J. van Andel and J. Aronson (eds) *Restoration Ecology*. Blackwell, Oxford, pp. 141–57. Covers many aspects of river and floodplain restoration and includes several examples of restoration projects.

More advanced texts

Calow, P. and Petts, G.E. (eds), 1996. *River Restoration*. Blackwell, Oxford. Contains chapters on flow allocations for in-river needs, environmentally sensitive river engineering, rehabilitation of river margins and channel restoration.

Downs, P.W. and Gregory, K.J., 2005. *River Channel Management: Towards Sustainable Hydrosystems*. Arnold, London. Written for higher-level undergraduates and postgraduates, this covers recent developments in river channel management.

Gilvear, D.J., 1999. Fluvial geomorphology and river engineering: future roles utilizing a fluvial hydrosystems framework. *Geomorphology*, 31: 229–45. Considers the present and future role of geomorphology in river management and river engineering.

Gordon, N.D., McMahon, T.A., Finlayson, B.L., Gippel, C.J. and Nathan, R.J., 2004. *Stream Hydrology: An Introduction for Ecologists*. John Wiley & Sons, Chichester. Further detail on ecological impacts and recent developments in channel management is provided in the final chapter.

Petts, G.E. and Amoros, C. (eds), 1996. *Fluvial Hydrosystems*. Chapman & Hall, London. Further information on ecological communities, links between geomorphological structure and ecology, human impacts and management.

Thorne, C.R., Hey, R.D. and Newson, M.D. (eds), 1997. *Applied Fluvial Geomorphology for River Engineering and Management*. John Wiley & Sons, Chichester. The chapter by Newson *et al.* provides some useful case studies to illustrate the application of geomorphology to river management.

Techniques

Hemphill, R.W., and Bramley, M.E., 1989. *Protection of River and Canal Banks*. Butterworth's, London. A useful guide to many different methods of providing bank protection. Also includes a detailed discussion of the various bank erosion processes.

River Restoration Centre, 1999. *River Restoration: Manual of Techniques: Restoring the River Cole and River Skerne, UK*. River Restoration Centre, Silsoe. Fully illustrated guide to the design and implementation of river restoration techniques used on the rivers Cole and Skerne.

Websites

The River Restoration Centre (UK), www.therrc.co.uk. The River Restoration Centre. Information on a number of demonstration projects in Denmark and the UK, lists of relevant publications, links to other websites and much more.

US Environmental Protection Agency River Corridor and Wetland Restoration, www.epa.gov/OWOW/restore/. Gives background on the principles of river restoration. Several case studies, numerous photographs and useful links.

Kissimmee River Restoration Project, www.sfwmd.gov/org/erd/krr/. Site giving general and specific information about this major restoration project in Florida. Includes several videos.

River Landscapes: restoring rivers and riparian lands, www.rivers.gov.au. An Australian site with definitions, fact sheets and examples of river restoration projects.

NOTES

3 The flow regime

[1] Potential evaporation is the rate of evaporation that would occur, given an unrestricted supply of water.

[2] It should be noted that the mean annual flood is *NOT* the same as the 1-in-1 year event.

5 Large-scale sediment transfer

[1] In this particular study effective precipitation was defined as the annual precipitation required to generate a given annual runoff at a standardised mean annual temperature of 50°F.

[2] A tributary fan is an accumulation of sediment deposited where a steep, sediment-laden tributary joins the main valley – see Chapter 8.

6 Flow in channels

[1] Since force = mass (in kg) × acceleration (in m s^{-2}), $1 N = 1$ kg m s^{-2}.

[2] Most of this friction is generated within the fluid itself as a result of turbulence.

[3] It should be noted that the depth of flow may also change in response to a change in width. See Box 6.2 for more information.

[4] **The molecular or dynamic viscosity** (represented by mu: μ) is measured in kg m^{-1} s^{-1} (kilograms per metre per second) or N s m^{-1} (newton seconds per metre). The molecular viscosity of water is 0.00114 kg m^{-1} s^{-1} at 15°C. **A kinematic viscosity** (v) was introduced by engineers to simplify the expression of viscosity in calculations: kinematic viscosity (v) = dynamic viscosity (μ) / density (p). Kinematic viscosity has units of m^2 s^{-1}, i.e. it has dimensions of length and time only.

7 Processes of erosion, transport and deposition

[1] Since work = force (in N) × distance (in metres), $1J = 1$ N m.

GLOSSARY OF TERMS

Abrasion   Mechanical wearing of rocks by sediment carried in the flow. This scratches, grinds, scours and polishes the rock surface.

Accelerated soil erosion   An increase in the rate of soil erosion brought about by human activity or natural processes. Net soil loss occurs if rates of soil erosion exceed rates of new soil formation.

Accretion   Accumulation of sediment in a particular location, such as a FLOODPLAIN or BAR surface.

Adjustable variable   See VARIABLE.

Aggradation   A net accumulation of sediment, which leads to an increase in the elevation of channel beds and FLOODPLAIN surfaces.

Allogenic change   Change in a system that is brought about by the influence of the external VARIABLES that control it.

Alluvial channel   A channel formed in ALLUVIUM.

Alluvial fan   A cone-shaped depositional feature that often form along mountain fronts, where steep, sediment-laden channels emerge from narrow valleys on to wide plains.

Alluvium   Sediment deposited by fluvial processes.

Anabranching channel   A type of river that flows in multiple channels, which are incised into the FLOODPLAIN, and which branch and rejoin around large islands.

Anastomosing channel   A sub-group of low-energy ANABRANCHING CHANNELS associated with AGGRADATION and relatively stable multi-channelled forms.

Annual flood   The highest flow at a given point on a river channel during a given year.

Armour layer   A layer of coarse sediment that forms in mixed gravel-bed rivers and which protects the finer underlying bed material. The particles may be tightly interlocked or IMBRICATED.

Attrition   The breaking down of a river's load into smaller fragments as a result of inter-particle collisions.

Autogenic change   Brought about when an internal THRESHOLD is crossed. For example, FLOODPLAIN stripping occurs in some environments when sediment accumulation increases the valley gradient to such an extent that rapid erosion results. The floodplain is then gradually built up again over time.

Avulsion   The abandonment of an existing channel when the flow switches to a new channel.

Backswamp   A low-lying marshy area that lies between the valley margin and the natural LEVEE of an alluvial channel.

Bankful discharge   The maximum flow that can be contained within the channel before it starts to inundate the floodplain.

Bar   An accumulation of sediment formed within a river channel. Rock bars are found in BEDROCK CHANNELS.

Base flow   The stable component of runoff which is sustained by GROUNDWATER FLOW, together with any runoff from lakes, wetlands, reservoirs and glacial melt.

Base level   The lowest point of a river's course, usually sea level. It marks the level below which a river cannot erode and is one of the EXTERNAL BASIN CONTROLS. **Local base levels** are imposed by various controls including lakes, reservoirs and resistant bands of rock, and control the depth of incision upstream from that point.

Bedforms   Morphological features that are formed on the channel bed by the action of flow over a deformable boundary.

Bedload   The coarser fraction of a river's load, which is moved by traction along the bed of the channel by rolling, sliding and SALTATION.

Bedrock channel   A channel formed directly in the underlying rock, rather than in ALLUVIUM.

Bed shear stress   A shearing force exerted by flowing water on the bed of the channel. See also SHIELDS PARAMETER.

Bernouilli effect   The reduction in pressure associated with an increase in flow velocity.

Boundary conditions   Define the boundaries within which a system operates. See also CHANNEL CONTROLS.

Boundary layer   The layer of flow closest to the channel boundary (or soil surface in the case of overland flow), through which a velocity gradient develops as a result of BOUNDARY RESISTANCE.

Boundary resistance   A type of flow resistance relating to the effects of the individual grains making up the channel boundary (**grain roughness**) and BEDFORMS (**form roughness**).

Boundary Reynolds number (Re*)   See GRAIN REYNOLDS NUMBER.

Braid bar accretion   A process of floodplain ACCRETION where abandoned braid bars become incorporated into FLOODPLAIN deposits.

Braided channel   A type of alluvial river channel in which the flow divides and rejoins around numerous bars. They are highly dynamic and are associated with rapid rates of channel migration.

Boulders   Bed particles that exceed 256 mm in diameter.

Capacity   Defines the volume of sediment that can be transported by a given flow. It decreases as the particle size becomes larger and increases with STREAM POWER.

Cascading system   The sub-system within which flows of energy and materials follow interconnected pathways through the fluvial system. Also called a **process system**.

Channel controls   Variables that control channel form, which include the FLOW REGIME and SEDIMENT REGIME (driving variables), and degree of VALLEY CONFINEMENT, CHANNEL SLOPE, CHANNEL SUBSTRATE and RIPARIAN VEGETATION (BOUNDARY CONDITIONS).

Channelisation   The artificial modification of river channels for flood control, navigation, land drainage and erosion control.

Channel network   The branching network of rivers and tributaries within a DRAINAGE BASIN.

Channel planform   The form of a river channel as viewed from above.

Channel resistance   A type of FLOW RESISTANCE relating to the three-dimensional shape of the river channel.

Channel slope   The amount of vertical drop in channel bed elevation along a given length of channel.

Channel substrate   The material that forms the channel boundary.

Clay particles   Very fine sediment with a diameter of less than 0.004 mm.

Cobbles   Bed particles that are intermediate in size between pebbles and boulders, with a diameter of 64 mm to 256 mm.

Cohesion   The term describing the ability of particles within a material to stick together.

Colluvium   Material that has been weathered and transported to the base of slopes by processes of mass movement and surface wash. It is distinct from ALLUVIUM, which has been deposited by fluvial processes.

Competence   Describes the ability of a particular flow to transport particles of a given size.

Complex response   Complex changes within a disturbed SYSTEM that are caused by feedbacks and interactions between the system VARIABLES.

Controlling variable   See VARIABLE.

Corrosion   The process of channel erosion whereby rock is dissolved and broken down by chemical action.

Crevasse splay   A fan-like depositional feature, formed when a LEVEE is breached and sediment-charged flow spreads out across the floodplain.

Critical flow   See FROUDE NUMBER.

Cross-sectional area   The area of channel flow viewed in cross-section.

Cut-off   An abandoned meander loop.

D$_{50}$ The median particle size in a sediment sample.

Degradation   1. Caused where net erosion leads to a reduction in the elevation of channel beds and floodplain surfaces. 2. Environmental degradation refers to a decline in the physical, chemical and biological condition of river channels, floodplains and wetland as a result of human activity.

Degrees of freedom   These reflect the ability of different components of the fluvial system to adjust.

Delta   A morphological feature formed when a river enters the relatively still waters of a lake or sea and deposits its load.

Deposition zone   The lower part of the drainage basin where deposition predominates. See also PRODUCTION ZONE and TRANSFER ZONE.

Discharge   The volume of flow passing a point in a given period of time.

Disequilibrium   A loss or lack of EQUILIBRIUM resulting in a state of instability.

Disequilibrium channel   See REGIME CHANNEL.

Drag force   The force exerted by a moving fluid on a stationary object. Moving objects, such as a grain of sediment falling through a still column of water, also experience drag.

Drainage basin   The area drained by a river and its tributaries. Also called the **catchment** or **watershed**.

Drainage density   The total length of stream channels within a drainage basin divided by the area of the basin.

Drainage divide   Defines the perimeter of a DRAINAGE BASIN. Also called the **catchment boundary**.

Driving force   See FORCE.

Drylands   The collective term for all types of arid and semi-arid regions.

Dynamic equilibrium   See EQUILIBRIUM.

Empirical   Based on observation or experiment, not on theory.

Energy   The capacity to do WORK.

Energy slope   See WATER SURFACE SLOPE.

Entrainment   The process by which particles are set in motion by flowing water as a result of FLUID FORCES.

Entrainment velocity   Defines the mean flow velocity at which particles of a given size will be set in motion by ENTRAINMENT.

Ephemeral   Refers to rivers which flow only part of the time and which may be dry for long periods.

Equilibrium   A system is in a state of **static equilibrium** when there is no change in it. In natural systems **steady state equilibrium** is more usual, where the system fluctuates around an unchanging mean. **Dynamic equilibrium** occurs when the system fluctuates around a changing mean.

Erodibility   Describes the susceptibility of natural materials to erosion. It is affected by characteristics such as sediment size, COHESION, slope length, vegetation and land use.

Erosivity   Describes the power of rainfall or flowing water to erode natural materials. It is affected by characteristics such as raindrop size and flow velocity.

External basin controls   The controlling variables of climate, TECTONICS, human activity and BASE LEVEL change that are external to the fluvial system.

External threshold   A transition within a SYSTEM from one state to another that is brought about by an external VARIABLE.

External variable   See VARIABLE.

Failure plane   See SHEAR PLANE.

Fall velocity   1. The velocity at which a particle in suspension falls through a column of still water. It varies with the volume, shape and density of the particle and the density and viscosity of the water. If the fall velocity is less than the upward component of turbulent velocity, the particle will remain aloft. 2. The mean flow velocity at which particles of a given size are deposited.

Feedback   A sequence of adjustments in response to a change in one of the system variables that either counteracts the effect of the original change (**negative feedback**) or enhances it (**positive feedback**).

Flood channel   Subsidiary channel found on floodplains and along valley margins that is only active during high flows.

Flood frequency   The frequency with which a flood of a given size can be expected to occur on average.

Floodplain   A relatively flat alluvial depositional landform that borders river channels and is periodically inundated by floodwater.

Flow regime   The 'climate' of a river, which is defined by average flow conditions over the course of a year.

Flow resistance   A general term for the forces that oppose the motion of water. These include BOUNDARY RESISTANCE, CHANNEL RESISTANCE and free-surface resistance associated with surface effects such as waves and hydraulic jumps.

Flow separation   Occurs when the LAMINAR SUBLAYER becomes detached from the channel boundary, for example where the flow moves around tight bends or over the crest of bedforms.

Fluid forces   The combined forces of LIFT and DRAG that are exerted on sediment grains at the bed of a channel.

Force   Anything which causes an object to move, or which changes the speed or direction of a moving object. Forces have magnitude and direction. In most situations in fluvial geomorphology several forces are involved, so the balance between **driving forces** and **resisting forces** is considered.

Form roughness   See BOUNDARY RESISTANCE.

Form system   See MORPHOLOGICAL SYSTEM.

Freeze–thaw action   A physical weathering process where the repeated growth and melting of ice crystals leads to the gradual enlargement of cracks and eventual disintegration of rocks.

Friction   A FORCE that resists relative motion between two bodies that are in contact.

Froude number (Fr)   The ratio between inertial and gravitational forces. In most channels, where gravitational forces dominate (Fr < 1), the flow is **subcritical**. A more rapid, streaming supercritical flow is seen when the inertial forces dominate (Fr > 1). **Critical flow** (Fr = 1) is transitional between these two flow types.

g   The acceleration due to gravity, which has an average value of 9.8 m s^{-2}.

Geomorphic unit   Landform-scale features such as channel BARS, RIFFLES AND POOLS, CUT-OFFS and MEANDER SCROLLS.

Gradually varied flow Associated with gradual changes in channel form which cause the flow to converge and diverge in a downstream direction. See also UNIFORM FLOW and RAPIDLY VARIED FLOW.

Grain Reynolds number (Re*) Combines the effects of near-bed velocity and grain size, allowing comparison to be made between the size of surface irregularities and the thickness of the LAMINAR SUBLAYER. Also known as the **boundary** or **shear Reynolds** number.

Grain roughness See FLOW RESISTANCE.

Gravel Sediment with a diameter of 2.0 mm to 16.0 mm.

Groundwater flow The subsurface movement of water below the water table in the saturated zone.

Gully A steep-sided, trench-like feature ranging in size from tens of centimetres to several metres deep. Commonly associated with DRYLAND environments but may also develop in response to poor land management practices.

Habitat The place where an organism lives, or the characteristic environment in which it is found.

Hillslope–channel coupling Defines the strength of the linkage between hillslopes and channels and is important in sediment transfer. Coupling is strong where steep hillslopes border a narrow valley, but becomes very weak where channels flow across wide floodplains.

Hydraulic drop A sudden decrease in flow depth caused by a transition from subcritical to supercritical flow. See also FROUDE NUMBER.

Hydraulic jump A breaking wave associated with a sudden increase in depth at which supercritical flow changes to subcritical flow. See also FROUDE NUMBER.

Hydraulic radius A measure of channel hydraulic efficiency defined by the ratio between cross-sectional area and WETTED PERIMETER.

Hydrograph A graph showing how DISCHARGE varies through time.

Imbrication A type of sediment structure where overlapping particles are inclined in the direction of flow.

Incision The downward cutting of a channel into its substrate.

Inertia The tendency of a stationary object to resist being moved or, in the case of moving objects, to resist a change in speed or direction.

Infiltration The process by which water enters the soil and moves down into it. The **infiltration capacity** is defined as the maximum rate at which a given soil can absorb water when in a given condition (Horton, 1933).

Internal threshold See THRESHOLD.

Internal variable See VARIABLE.

Inter-rill erosion Erosion of the soil surface between RILLS by rain splash and sheet erosion.

Isovel A line joining points of equal velocity.

Kinetic energy The energy possessed by moving objects.

Knickpoint A break in slope in the LONG PROFILE of a river which is marked by a waterfall or series of rapids.

Lag deposit The deposition of relatively coarse grains on the channel bed as a result of selective transport.

Laminar flow A type of flow observed in viscous fluids and very low-velocity flows of water, where the fluid moves smoothly as a series of thin layers. See also REYNOLDS NUMBER.

Laminar sublayer A thin layer immediately adjacent to the bed of a channel where the velocity is greatly reduced by boundary resistance and flow is LAMINAR.

Lateral accretion A process of floodplain ACCRETION whereby material is deposited as a channel migrates laterally across its FLOODPLAIN.

Levee Raised ridges that run along channel margins which are formed by the deposition of relatively coarse SUSPENDED material during OVERBANK flows. **Artificial levees** are constructed for flood defence to increase channel capacity at high flows.

Lift force A FLUID FORCE caused by a drop in pressure above a grain of sediment that is brought about by the BERNOUILLI EFFECT.

Lithology Refers to rock characteristics such as grain size, and physical and chemical properties.

Long profile Describes the gradient of the water surface along a river, from source to mouth. The typical shape is 'concave to the sky', although profiles can also be straight or convex, and irregularities are often seen.

Manning's 'n' An empirically derived roughness coefficient. See also RESISTANCE EQUATIONS.

Mass movement A down-slope movement of bulk sediment as landslides, slips, slumps and debris flows.

Mass wasting See MASS MOVEMENT.

Meander scrolls Deposits formed by the migration of meanders across the FLOODPLAIN which are laid down to produce concentric **ridges** separated by lower elevation troughs, or **swales**.

Meander wavelength The straight-line distance from the apex of one meander loop to the apex of the next loop (on the same side of the valley).

Meandering channel A single channel that follows a winding course, with a SINUOSITY RATIO of 1.5 or more.

Momentum Relates to the energy possessed by a moving object and is dependent on the mass of the object and its velocity.

Morphological system The sub-system of morphological components. Also called a **form system**.

Negative feedback See FEEDBACK.

Non-regime channel See REGIME CHANNEL.

Overbank flow Occurs when the channel capacity is exceeded and flow proceeds to inundate the surrounding FLOODPLAIN.

Overland flow A flow of water over the land surface as a result of saturated conditions (**saturation overland flow**), or where rainfall intensity exceeds the INFILTRATION CAPACITY (**Hortonian overland flow**).

Oxbow lake A small crescent-shaped lake that occupies a CUT-OFF meander bend.

Palaeochannel An abandoned channel preserved within the FLOODPLAIN deposits.

Perennial Refers to rivers that flow year-round.

Point bar A channel BAR that forms on the inside of a meander bend.

Pore water pressure The pressure of the water filling pores and crevices in soils, REGOLITH and bank materials. When all the pores are filled with water, a positive pore water pressure develops (greater than atmospheric). When a moisture deficit exists, the remaining water is held in pores by suction forces and a negative pore water pressure develops (less than atmospheric).

Positive feedback See FEEDBACK.

Potential energy The energy stored by water and rocks that are elevated above sea level.

Pothole A deep circular scour feature found in bedrock channels.

Pressure head Refers to the pressure exerted by a depth of fluid on the bed of a channel.

Primary erosion The initial erosion of material that has not previously been eroded and transported.

Process–response system A system in which the processes of the CASCADING SYSTEM interact with the forms of the MORPHOLOGICAL SYSTEM.

Process system See CASCADING SYSTEM.

Production zone The headwater regions of the drainage basin where flow and sediment are generated. See also TRANSFER ZONE and DEPOSITION ZONE.

Pulsed disturbance An episodic disturbance of low frequency and high magnitude such as a major flood event.

Ramped disturbance A system disturbance caused by a long-term change in one of the controlling variables.

Rapidly varied flow Associated with sudden changes in channel geometry, such as constrictions and expansions, which lead to HYDRAULIC JUMPS and DROPS.

Reach A length of river along which the CHANNEL CONTROLS are sufficiently uniform to allow a fairly consistent morphological structure to be maintained.

Reaction time The time period over which the impact of a change is absorbed by a SYSTEM.

Realignment The artificial straightening of river channels for navigation and flood control.

Regime channel A given channel is 'in regime' when its morphological characteristics, such as size, fluctuate around a mean condition over the time scale considered. This condition is not true for **non-regime**, or **disequilibrium**, channels.

Regolith The term for *in situ* weathered material, which can vary in size from large boulders to fine clays and microscopic colloids.

Relaxation time A period of adjustment that occurs as a SYSTEM moves towards a new EQUILIBRIUM after a disturbance. It is preceded by a REACTION TIME.

Re-sectioning Refers to the artificial modification of the channel cross-section.

Resistance The power of an object or material to resist a force that is imposed on it. See also FLOW RESISTANCE.

Resistance equations Empirical equations that relate mean flow velocity, channel slope, HYDRAULIC RADIUS and FLOW RESISTANCE.

Return period The frequency, in years, with which a given event can be expected to occur. See also FLOOD FREQUENCY.

Reynolds number (Re) A dimensionless coefficient which expresses the ratio between inertial and viscous forces acting on a body of fluid. At Re numbers of less than 500 viscous forces dominate and the flow is LAMINAR; when Re exceeds 2,100 the flow becomes TURBULENT; transitional flow is observed between Re values of 500 and 2,100. See also GRAIN REYNOLDS NUMBER.

Ridge and swale topography See MEANDER SCROLLS.

Riffles and pools Periodic undulations in bed elevation where relatively shallow coarse-grained riffles are separated by deeper pools with a spacing of between five and seven times the channel width.

Rill A small channel that forms on slopes as a result of concentrated OVERLAND FLOW.

Riparian On, or relating to, the banks of a river.

Riparian vegetation The vegetation that grows within RIPARIAN ZONES.

Riparian zones The term applied to the belts of land that run alongside the banks of a river.

River restoration The implementation of techniques to restore the natural function of a river channel that has become degraded (in physical, chemical and biological terms) as a result of human activity.

River training The installation of artificial flow structures to encourage scour or deposition within a river channel.

Runoff The movement of water above and below the surface.

Saltation A type of bedload transport where particles are moved along the bed in a series of short jumps.

Sand grains Particles with a diameter of 0.062 mm to 2.0 mm.

Secondary flows Currents that move across the channel, perpendicular to the main direction of flow.

Sediment budget An organising framework to quantify inputs, transfers, stores and outputs of sediment from a geomorphological system.

Sediment delivery ratio (SDR) The proportion of sediment produced by primary erosion that actually exits the drainage basin.

Sediment regime The supply of sediment to a channel reach defined in terms of its volume and size.

Sediment store A depositional environment in which sediment is stored. Storage is temporary if the sediment is re-eroded at a later stage.

Sediment yield   The total amount of sediment that exits a drainage basin, or passes a given point, over a period of time. It includes BEDLOAD and SUSPENDED LOAD. See also SPECIFIC SEDIMENT YIELD.

Shear plane   A curved or planar surface within bank or slope materials across which the applied SHEAR STRESS is equal to the SHEAR STRENGTH of the material.

Shear strength   The strength of materials to resist deformation when a SHEAR STRESS is applied.

Shear stress   A shearing force that acts tangentially to the surface of an object. See also BED SHEAR STRESS.

Shear velocity   A measure of the vertical velocity gradient and shear stress.

Shields parameter   The critical bed shear stress required to entrain a particle of sediment. It is expressed in dimensionless terms.

Silt   Fine particles, intermediate in size between fine sands and clays, with a diameter of 0.004 mm to 0.062 mm.

Sinuosity ratio   A measure of the degree of meandering along a section of channel given by the ratio of the channel length and the valley length.

Solute load   Dissolved material carried in the flow as ions.

Sorting   The grading of sediment by flowing water. Sediment of a similar size tends to be deposited at the same time because the FALL VELOCITY is determined by grain size. Poorly sorted sediments consist of a mixed range of particle sizes.

Specific sediment yield   The total amount of suspended and bedload sediment that exits a catchment in a given time. It is usually measured in tonnes per year per unit area of the drainage basin.

Specific stream power   The stream power per unit area of the bed.

Stage   The height of the water level in a river.

Static equilibrium   See EQUILIBRIUM.

Steady flow   The term used to describe channel flow which has a constant discharge over time. See also UNSTEADY FLOW.

Steady-state equilibrium   See EQUILIBRIUM.

Steps and pools   Sequences of steps formed from coarse material separated by relatively still pools. Found in steep, upland channel reaches.

Stream power   The rate at which WORK is carried out along a given length of channel, stream power increases with channel slope and discharge.

Subcritical flow   See FROUDE NUMBER.

Supercritical flow   See FROUDE NUMBER.

Supply limited   Sediment transport rates that are limited by the supply of sediment, rather than the ability of the flow to transport it.

Suspended load   Fine sediment that is transported within the turbulent flow profile and held aloft by eddies.

System   A collection of objects and the processes that link those objects together.

Tectonics   Refers to the deformation of the Earth's crust as a result of internal forces.

Terraces   Remnants of former FLOODPLAIN surfaces that have been incised by the channel to create a new floodplain at a lower elevation. Terraces are gently sloping surfaces found along valley margins, and are separated from the main valley floor by a steep scarp slope.

Thalweg   The area of maximum water velocity within a channel flow.

Threshold   A critical condition where a SYSTEM changes from one state of operation to another as a result of a change in one of the controls. **External thresholds** are crossed as a result of change in one of the external VARIABLES. **Internal thresholds** are crossed as a result of instability developed within the system.

Throughflow   A hydrological pathway by which water travels downslope through the soil and parallel to the soil surface. Also referred to as **interflow**.

Transfer zone   The zone in the central part of a drainage basin, through which sediment is transported from the PRODUCTION ZONE to the DEPOSITION ZONE.

Transmission loss   The downstream loss of flow from a channel as a result of seepage through the channel boundary and/or high evaporation rates.

Transport-limited   Sediment transport rates that are limited by the CAPACITY or COMPETENCE of the flow, rather than the supply of sediment.

Turbulent flow   The type of flow that prevails when inertial forces are greater than the viscous forces. It is characterised by eddies, mostly generated in the near-bed region, which transfer momentum throughout the flow. See also REYNOLDS NUMBER.

Uniform flow   Describes flow in straight channels with a uniform cross-section, where the depth and mean flow velocity remain constant along the channel. See also GRADUALLY VARIED FLOW and RAPIDLY VARIED FLOW.

Unsteady flow   The term used to describe channel flow with a variable discharge. The flow in natural river channels is unsteady. See also STEADY FLOW.

Valley confinement   The degree to which a river channel is confined between the valley walls.

Variables   Quantities whose value changes though time. **Internal variables** operate within a given system. **External variables** are controlled by forces outside the system.

A **controlling variable** controls the adjustment of another variable, which is called the **adjustable variable**.

Vertical accretion   The vertical build up of fine sediment that is deposited on the floodplain surface during overbank flows.

Viscosity   The resistance of a fluid to deformation and flow.

Wandering channel   A type of anabranching channel that is transitional between braided and meandering.

Wash load   The finest particles (less than 0.0063 mm in diameter) within the suspended load.

Water surface slope   The downstream change in water surface elevation along a length of channel. This closely approximates the **energy slope**, the steepness of which reflects the rate at which POTENTIAL ENERGY is being converted to KINETIC ENERGY.

Weathering   The chemical decomposition and physical breakdown of rocks *in situ* at the surface of the Earth. Transport is not involved.

Wetted perimeter   The cross-sectional length of a channel boundary that is in contact with the flow.

Work   Carried out when a FORCE moves an object.

REFERENCES

Alexander, J., Bridge, J., Leeder, M.R., Collier, R.E.L. and Gawthorpe, R.L., 1994. Holocene meander-belt evolution in an active extensional basin, southwestern Montana. *Journal of Sedimentary Research*, B64: 542–59.

Andrews, E.D., 1980. Effective and bankfull discharges of streams in the Yampa River basin, Colorado and Wyoming. *Journal of Hydrology*, 46: 311–30.

Andrews, E.D., 1983. Entrainment of gravel from naturally sorted riverbeds material. *Geological Society of America Bulletin*, 94: 1225–31.

Ashmore, P.E., 1991. How do gravel bed rivers braid? *Canadian Journal of Earth Sciences*, 28: 326–41.

Ashworth, P.J. and Ferguson, R.I., 1986. Interrelationships of channel processes, changes and sediments in a proglacial braided river. *Geografisker Annaler*, 68A: 361–71.

Ashworth, P.J. and Ferguson, R.I., 1989. Size selective entrainment of bedload in gravel bed streams. *Water Resources Research*, 25: 627–34.

Bagnold, R.A., 1960. *Some Aspects of the Shape of River Meanders*, United States Geological Survey Professional Paper 282E.

Bagnold, R.A., 1980. An empirical correlation of bedload transport rates in flumes and natural rivers. *Proceedings of the Royal Society of London, Series A*, 372: 453–73.

Baker, V.R., 1973. Palaeohydrology and sedimentology of Lake Missoula flooding in eastern Washington. *Geological Society of America Special Paper* 144.

Baker, V.R. and Pickup, G., 1987. Flood geomorphology of the Katherine Gorge, Northern Territory, Australia. *Bulletin of the Geological Society of America*, 98: 635–46.

Bakoariniaina, L.N., Kusky, T. and Raharimahefa, T., 2006. Disappearing Lake Alaotra: monitoring catastrophic erosion, waterway silting, and land degradation hazards in Madagascar using Landsat imagery. *Journal of African Earth Sciences*, 44: 241–52.

Bathurst, J.C., 1987a. Measuring and modelling bedload transport in channels with coarse bed materials. In: K.S. Richards (ed.), *River Channels: Environment and Process*. Blackwell, Oxford, pp. 272–94.

Bathurst, J.C., 1987b. Critical conditions for bed material movement in steep boulder-bed streams. In: *Erosion and Sedimentation in the Pacific Rim*. IAHS Publication 165, Institute of Hydrology, Wallingford, pp. 309–18.

Bathurst, J.C., 1993. Flow resistance through the channel network. In: K. Beven and M.J. Kirkby (eds),

Channel Network Hydrology. John Wiley & Sons, Chichester, pp. 43–68.

Beckinsale, R.P., 1969. River regimes. In: R.J. Chorley (ed.), *Introduction to Physical Hydrology*. Methuen, London, pp. 176–92.

Bell, M. and Walker, M.J.C., 2005. *Late Quaternary Environmental Change: Physical and Human Perspectives*. Second edition. Prentice Hall, Harlow.

Belsky, A.J., Matzke, A. and Uselman, S., 1999. Survey of livestock influences on stream and riparian ecosystems in the western United States. *Journal of Soil and Water Conservation*, 54(1): 419–31.

Benda, L. and Dunne, T., 1997. Stochastic forcing of sediment supply to channel networks from landsliding and debris flow. *Water Resources Research*, 33(12): 2849–63.

Benito, G., Lang, M., Barriendos, M., Llasat, M.C., Francés, F., Ouarda, T., Thorndycraft, V., Enzel, Y., Bardossy, A., Coeur, D. and Bobée, B., 2004. Use of systematic, palaeoflood and historical data for the improvement of flood risk estimation: review of scientific methods. *Natural Hazards*, 31: 623–43.

Beven, K., 1981. The effect of ordering on the geomorphological effectiveness of hydrologic events. *IAHS-AISH Publication*, 132: 510–26.

Biron, P.M., Lane, S.N., Roy, A.G., Bradbrook, K. and Richards, K.S., 1998. Sensitivity of bed shear stresses estimated from vertical velocity profiles: the problem of sampling resolution. *Earth Surface Processes and Landforms*, 23: 133–9.

Bluck, B.J., 1982. Texture of gravel bars in braided streams. In: R.D. Hey, J.C. Bathurst and C.R. Thorne (eds) *Gravel-bed Rivers*, John Wiley & Sons, Chichester, pp. 339–55.

Boardman, J., Foster, I.D.L. and Dearing, J.A. (eds), 1990. *Soil Erosion on Agricultural Land*. John Wiley & Sons, Chichester.

Bogen, J., 1995. Sediment transport and deposition in mountain rivers. In: I.D.L., Foster, A.M. Gurnell and B.W. Webb (eds), *Sediment and Water Quality in River Catchments*. John Wiley & Sons, Chichester, pp. 437–51.

Bravard, J.-P. and Petts, G.E., 1996. Human impacts on fluvial systems. In: G.E. Petts and C. Amoros (eds), *Fluvial Hydrosystems*. Chapman & Hall, London, pp. 242–62.

Brayshaw, A.C., 1984. The characteristics and origin of cluster bedforms in coarse-grained alluvial channels. In: C.H. Koster and R.H. Stell (eds), *Sedimentology of Gravels and Conglomerates*. Canadian Society of Petroleum Geologists Memoir 10, 77–85.

Bridge, J.S., 1993. The interaction between channel geometry, water flow, sediment transport and deposition in braided rivers. In: J.L. Best and C.S. Bristow (eds), *Braided Rivers*. Special Publication of the Geological Society of London 75, pp. 13–71, London.

Bridge, J.S., 2003. *Rivers and Floodplains: Forms, Processes, and Sedimentary Record*. Blackwell, Oxford.

Brierley, G.J. and Fryirs, K.A., 2005. *Geomorphology and River Management: Applications of the River Styles Framework*. Blackwell, Oxford.

Brookes, A., 1985. River channelization: traditional engineering methods, physical consequences and alternative practices. *Progress in Physical Geography*, 9: 44–73.

Brookes, A., 1988. *Channelized Rivers: Perspectives for Environmental Management*. J. Wiley & Sons, Chichester.

Brooks, S.M. and Richards, K.S., 1994. The significance of rainstorm variation to shallow translational failure. *Earth Surface Processes and Land-forms*, 19: 85–94.

Brunsden, D. and Thornes, J.B., 1979. Landscape sensitivity and change. *Transactions of the Institute of British Geographers*, NS 4: 463–84.

Bull, L.J. and Kirkby, M.J. (eds), 2002. *Dryland Rivers: Hydrology and Geomorphology of Semi-arid Channels*. John Wiley & Sons, Chichester.

Bull, W.B., 1991. *Geomorphic Responses to Climatic Change*. Oxford University Press, New York.

Bunn, S.E. and Arthington, A.H., 2002. Basic principles and ecological consequences of altered flow regimes for aquatic biodiversity. *Environmental Management*, 30(4): 492–507.

Burnett, A.W. and Schumm, S.A., 1983. Alluvial river response to neotectonic deformation in Louisiana and Mississippi. *Science*, 222: 49–50.

Carling, P., 1988. The concept of dominant discharge applied to two gravel-bed streams in relation to channel stability thresholds. *Earth Surface Processes and Landforms*, 13: 355–67.

Carling, P.A., 1991. An appraisal of the velocity-reversal hypothesis for stable pool–riffle sequences in the River Severn. *Earth Surface Processes and Landforms*, 16: 19–31.

Carson, M.A. and Kirkby, M.J., 1972. *Hillslope Form and Process*. Cambridge University Press, Cambridge.

Chanson, H., 1999. *The Hydraulics of Open Channel Flow: An Introduction*. Arnold, London.

Charlton, M.E., Large, A.R.G. and Fuller, I.C., 2003. Application of airborne LIDAR in river environments: the River Coquet, Northumberland, UK. *Earth Surface Processes and Landforms*, 28: 299–306.

Chin, A., 1999. The morphologic structure of step-pools in mountain streams. *Geomorphology*, 27: 191–204.

Chorley, R.J., Schumm, S.A. and Sugden, D.E., 1984. *Geomorphology*. Methuen, London and New York.

Chow, V.T., 1959. *Open Channel Hydraulics*. McGraw-Hill, Kogakusha, Tokyo.

Church, M., 1983. Pattern of instability in a wandering gravel bed channel. In: J.D. Collinson and J. Lewin (eds), *Modern and Ancient Fluvial Systems*, Special Publication of the International Association of Sedimentologists, 6. Blackwell, Oxford, pp. 169–80.

Church, M., 1992. Channel morphology and typology. In: P. Calow and G.E. Petts (eds), *The River Handbook*, I Blackwell, Oxford, pp. 126–43.

Church, M. and Jones, D., 1982. Channel bars in gravel bed rivers. In: R.D. Hey, J.C. Bathurst and C.R. Thorne, (eds), *Gravel Bed Rivers*. John Wiley & Sons, Chichester, pp. 291–338.

Church, M. and Miles, M.J., 1982. Processes and mechanisms of bank erosion. In: R.D. Hey, J.C. Bathurst and C.R. Thorne (eds), *Gravel Bed Rivers*. John Wiley & Sons, Chichester, pp. 221–71.

Church, M. and Slaymaker, O., 1989. Disequilibrium of Holocene sediment yield in glaciated British Columbia. *Nature*, 337: 452–4.

Closs, G., Downes, B. and Boulton, A., 2004. *Freshwater Ecology*. Blackwell, Oxford.

Collier, M.P., Webb, R.H. and Andrews, E.D., 1997. Experimental flooding in Grand Canyon. *Scientific American*, 276(1): 82–9.

Collins, A.L., Walling, D.E. and Leeks, G.J.L., 1997. Use of the geochemical record preserved in floodplain deposits to reconstruct recent changes in river basin sediment sources. *Geomorphology*, 19: 151–67.

Cooke, R.U. and Doornkamp, J.C., 1990. *Geomorphology in Environmental Management: A New Introduction*. Clarendon Press, Oxford.

Costa, J.E., 1987. Hydraulics and basin morphometry of the largest flash floods in the conterminous United States. *Journal of Hydrology*, 93: 313–38.

Coulthard, T.J., 2001. Landscape evolution models: a software review. *Hydrological Processes*, 15: 165–73.

Coulthard, T.J., Lewin, J. and Macklin, M.G., 2005. Modelling differential catchment response to environmental change. *Geomorphology*, 69(1–4): 222–41.

Curtis, W.F., Culbertson, J.K. and Chase, E.B., 1973. Fluvial-sediment discharge to the oceans from the conterminous United States. *US Geological Survey Circular 670*, United States Geological Survey.

Dietrich, W.E., 1987. Mechanics of flow and sediment transport in river bends. In: K.S. Richards (ed.) *River Channels: Environment and Process*. Blackwell, Oxford, pp. 179–227.

Dietrich, W. E., and Dunne, T., 1978. Sediment budget for a small catchment in mountainous terrain. *Zeitschrift für Geomorphologie, Suppl.* 29: 191–206.

Dietrich, W.E., Smith, J.D. and Dunne, T., 1979. Flow and sediment transport in a sand-bedded meander. *Journal of Geology*, 87: 305–15.

Dietrich, W.E., Kirchener, J.W., Ikeda, H. and Iseya, F., 1989. Sediment supply and the development of the coarse surface layer in gravel-bedded rivers. *Nature*, 340: 215–17.

Downs, P.W. and Gregory, K.J., 2004. *River Channel Management: Towards Sustainable Catchment Hydrosystems*. Arnold, London.

Dunkerley, D.L., 1992. Channel geometry, bed material and inferred flow conditions in ephemeral stream systems, barrier range, western N.S.W., Australia. *Hydrological Processes*, 6(4): 417–33.

Dunne, T., 1978. Field studies of hillslope processes. In: M.J. Kirkby (ed.) *Hillslope Hydrology*. John Wiley & Sons, Chichester, pp. 227–94.

Dunne, T. and Black, R.D., 1970. Partial area contributions to storm runoff in a small New England watershed. *Water Resources Research*, 6: 1296–311.

Dury, G.H., 1977. Underfit streams: retrospect, perspect and prospect. In: K.J. Gregory (ed.), *River Channel Changes*. John Wiley & Sons, Chichester, pp. 281–93.

DWAF, 2001. *The Working for Water Programme: Annual Report 2000/1*, Department of Water Affairs and Forestry, Pretoria, South Africa.

Emmett, W.W., 1980. *A Field Calibration of the Sediment-trapping Characteristics of the Helley–Smith Bedload Sampler*. United States Geological Survey Professional Paper 1139.

Erskine, W.D., 1996. Response and recovery of a sand-bed stream to a catastrophic flood. *Zeitschrift für Geomorphologie*, 40: 359–83.

Erskine, W.D. and Livingstone, E.A., 1999. In-channel benches: the role of floods in their formation and destruction on bedrock-confined rivers. In: A.J. Miller and A. Gupta (eds), *Varieties of Fluvial Form*. John Wiley & Sons, Chichester, pp. 445–76.

Erskine, W.D. and Warner, R.F., 1988. Geomorphic effects of alternating flood-and-drought-dominated regimes on NSW coastal rivers. In: R.F. Warner (ed.), *Fluvial Geomorphology of Australia*. Academic Press, Sydney, pp. 303–22.

Erskine, W.D. and Warner, R.F., 1999. Significance of river bank erosion as a sediment source in the alternating flood regimes of southeastern Australia. In: A. Brown and T. Quine (eds), *Fluvial Geomorphology and Environmental Change*. John Wiley & Sons, Chichester, pp. 139–63.

Farquharson, F.A.K., Meigh, J.R. and Sutcliffe, J.V., 1992. Regional flood frequency analysis in arid and semi-arid areas. *Journal of Hydrology*, 138(3–4): 287–501.

Ferguson, R.I., 1979. River meanders: regular or random? In: N. Wrigley (ed.), *Statistical Applications in the Spatial Sciences*. Pion, London, pp. 229–41.

Ferguson, R.I., 1981. Channel form and channel changes. In: J. Lewin (ed.), *British Rivers*. Allen & Unwin, London, pp. 90–125.

Ferguson, R.I., 1987. Hydraulic and sedimentary controls of channel pattern. In: K.S. Richards (ed.), *River Channels: Environment and Process*. Blackwell, Oxford, pp. 129–58.

Ferguson, R.I., 1993. Understanding braiding processes in gravel-bed rivers: progress and unsolved problems. In: J.L. Best and C.S. Bristow (eds), *Braided Rivers*. Special Publication of the Geological Society of London 75, pp. 73–87.

Foster, I.D.L., Owens, P.N. and Walling, D.E., 1996. Sediment yields and sediment delivery in the catchments of Slapton Lower Ley, South Devon, UK. *Field Studies*, 8: 629–61.

Gilbert, G.K., 1917. *Hydraulic-Mining Debris in the Sierra Nevada*. United States Geological Survey Professional Paper. 105.

Gilvear, D.J., 1999. Fluvial geomorphology and river engineering: future roles utilizing a fluvial hydrosystems framework. *Geomorphology*, 31: 229–45.

Gilvear, D.J. and Bryant, R., 2003. Analysis of aerial photography and other remotely sensed data. In: G.M. Kondolf and H. Piégay (eds), *Tools in Fluvial Geomorphology*. John Wiley & Sons, Chichester, pp. 135–70.

Gilvear, D.J., Winterbottom, S.J. and Sichingabula, H., 2000. Mechanisms of channel planform change on the actively meandering Luangwa river, Zambia. *Earth Surface Processes and Landforms*, 25: 421–36.

Gomez, B. and Church, M., 1989. An assessment of bedload sediment transport formulae for gravel bed rivers. *Water Resources Research*, 25: 1161–86.

Gordon, N.D., McMahon, T.A., Finlayson, B.L., Gippel, C.J. and Nathan, R.J., 2004. *Stream Hydrology: An Introduction for Ecologists*. John Wiley & Sons, Chichester.

Goudie A., 1981. *Geomorphological Techniques*. Allen & Unwin, London.

Graf, W.L., 1975. The impact of suburbanisation on fluvial geomorphology. *Water Resources Research*, 11: 690–2.

Gregory, K.J., 1995. Human activity and palaeohydrology. In: K.J. Gregory, L. Starkel and V.R. Baker (eds), *Global Continental Palaeohydrology*. John Wiley & Sons, Chichester, pp. 151–72.

Hall M.J., 1970. A critique of methods of simulating rainfall. *Water Resources Research*, 6: 1104–14.

Hancock, G.S., Anderson, R.S. and Whipple, K.X., 1998. Beyond power: bedrock river incision process and form. In: K.J. Tinkler and E.E. Wohl (eds), *Rivers over Rock: Fluvial Processes in Bedrock Channels*. Geophysical Monograph Series. American Geophysical Union, Washington DC: 35–60.

Harbor, D.J., Schumm, S.A. and Harvey, M.D., 1994. Tectonic control of the Indus River in Sindh, Pakistan. In: S.S. Schumm and B.R. Winkley (eds), *The Variability of Large Alluvial Rivers*. American Association of Civil Engineers Press, New York, pp. 161–76.

Harvey, A., 1982. The role of piping in the development of badlands and gully systems in south-east Spain. In: R. Bryan and A. Yair (eds), *Badland: Geomorphology and Piping*. Geobooks, Cambridge, 317–35.

Harvey, M.D. and Schumm, S.A., 1994. Alabama River: variability of overbank flooding and deposition. In: S.A. Schumm and B.R. Winkley (eds), *The Variability of Large Alluvial Rivers*, American Society of Civil Engineers, New York, pp. 313–37.

Helley, E.J. and Smith, W., 1971. *Development and Calibration of a Pressure-difference Bedload Sampler*. United States Geological Survey Open File Report 18, United States Geological Survey.

Heritage, G.L. and Hetherington, D., 2007. Towards a protocol for laser scanning in fluvial geomorphology. *Earth Surface Processes and Landforms*, 32: 66–74.

Heritage, G.L., Broadhurst, L.J. and Birkhead, A.L., 2001. The influence of contemporary flow regime on the geomorphology of the Sabie River, South Africa. *Geomorphology*, 38: 197–211.

Heritage, G.L., van Niekerk, A.W. and Moon, B.P., 1999. Geomorphology of the Sabie River, South Africa: an incised bedrock-influenced channel. In: A.J. Miller and A. Gupta (eds), *Varieties of Fluvial Form*. John Wiley & Sons, Chichester, pp. 33–52.

Heritage, G.L., Large, A.R.G., Moon, B.P. and Jewitt, G., 2004. Channel hydraulics and geomorphic effects of an extreme flood event on the Sabie River, South Africa. *Catena*, 58: 151–81.

Hewlett, J.D. and Hibbert, A.R., 1967. Factors affecting the response of small watersheds to precipitation in humid areas. *Proceedings of the International Symposium on Forest Hydrology* (1965), *Pennsylvania State University*, Pergamon, New York, 275–90.

Hey, R.D., 1988. Bar form resistance in gravel-bed rivers. *Journal of Hydraulic Engineering*, 114: 1498–508.

Hey, R.D., 1994. Environmentally sensitive river engineering. In: P. Calow and G.E. Petts (eds), *The Rivers Handbook: Hydrological and Ecological Principles*. Blackwell, Oxford, pp. 337–62.

Hey, R.D. and Thorne, C.R., 1986. Stable channels with mobile gravel beds. *Journal of Hydraulic Engineering*, 112: 671–89.

Hickin, E.J. and Nanson, G., 1975. The character of channel migration on the Beatton River, north-east British Columbia, Canada. *Bulletin of the Geological Society of America*, 86: 487–94.

Hickin, E.J. and Nanson, G., 1984. Lateral migration rates of river bends. *Journal of Hydraulic Engineering*, 110: 1557–67.

Hjulstrøm, F., 1935. Studies of the morphological activity of rivers as illustrated by the River Fryis. *Bulletin of the Geological Institute of Uppsala*, 25: 221–527.

Hollis, G.E., 1975. The effect of urbanisation on floods of different recurrence intervals. *Water Resources Research*, 11: 431–4.

Hooke, J.M., 2003. Coarse sediment connectivity in river channel systems: a conceptual framework and methodology. *Geomorphology*, 56: 79–94.

Hooke, J.M. and Harvey, A.M., 1983. Meander changes in relation to bend morphology and secondary flows. In: J. Collinson and J. Lewin (eds), *Modern and Ancient Fluvial Systems*. Special Publication of the International Association of Sedimentologists, 6. Blackwell, Oxford, pp. 121–32.

Horton, R.E., 1933. The role of infiltration in the hydrologic cycle. *Transactions of the American Geophysical Union*, 14: 446–60.

Horton, R.E., 1945. Erosional development of streams and their drainage basins: hydrophysical approach to quantitative morphology. *Bulletin of the Geological Society of America*, 56: 275–370.

Howard, A.D., 1994. A detachment limited model of drainage basin evolution. *Water Resources Research*, 30(7): 2261–85.

Huggett, R.J., 2003. *Fundamentals of Geomorphology*. Routledge, London.

International Commission on Large Dams, 1988. *World Register of Large Dams: Update*. ICOLD, Paris.

James, A., 1999. Time and the persistence of alluvium: river engineering, fluvial geomorphology, and mining sediment in California. *Geomorphology*, 31: 265–90.

Jiongxin, X. and Yunxia, Y., 2005. Scale effects on specific sediment yield in the Yellow River basin and geomorphological explanations. *Journal of Hydrology*, 307: 219–32.

Jones, J.A.A., 1979. Extending the Hewlett model of stream runoff generation. *Area*, 11(2): 110–14.

Jones, J.A.A., 1997. *Global Hydrology*. Longman, Harlow.

Jorgensen, D.W., Harvey, M.D., Schumm, S.A. and Flam, L., 1993. Morphology and dynamics of the Indus River: implications for the Mohenjo Daro site. In: J.F. Shroeder, Jr (ed.), *Himalaya to the Sea*. Routledge, London, pp. 288–326.

Kay, M., 1998. *Practical Hydraulics*. E. & F.N. Spon, London.

Keller, E.A., 1972. Development of alluvial stream channels: a five stage model. *Geological Society of America Bulletin*, 83: 1531–6.

Keller, E.A. and Melhorn, N., 1978. Rhythmic spacing and origin of pools and riffles. *Geological Society of America Bulletin*, 89: 723–30.

Kingsford, R.T. and Thomas, R.F., 1995. The Macquarie Marshes in arid Australia and their waterbirds: a 50 year history of decline. *Environmental Management*, 19: 867–78.

Kirkby, M.J. and Chorley, R.J., 1967. Throughflow, overland flow and erosion. *Bulletin of the International Association of Scientific Hydrology*, 12: 5–21.

Knight, D.W., 1989. Hydraulics of flood channels. In: K. Beven and P. Carling (eds), *Floods: Sedimentological and Geomorphological Implications*. John Wiley & Sons, Chichester, pp. 83–105.

Knight, D.W. and Shiono, K., 1996. River channel and floodplain hydraulics. In: M.G. Anderson, D.E. Walling and P.D. Bates (eds), *Floodplain Processes*. John Wiley & Sons, Chichester, pp. 139–81.

Knighton, D.A., 1998. *Fluvial Forms and Processes: A New Perspective*. Arnold, London.

Knighton, D. and Nanson, G., 1997. Distinctiveness, diversity and uniqueness in arid zone river systems. In: D.S.G. Thomas (ed.), *Arid Zone Geomorphology*. John Wiley & Sons, Chichester, pp. 185–203.

Knox, J.C., 1995. Fluvial systems since 20,000 years BP. In: K.J. Gregory and L. Starkel (eds), *Global*

Continental Palaeohydrology. John Wiley & Sons, Chichester, pp. 87–108.

Kondolf, G.M. and Swanson, M.L., 1993. Channel adjustments to reservoir construction and gravel extraction along Stony Creek, California. *Environmental Geology*, 21: 256–69.

Kozarski, S. and Rotnicki, K., 1977. Valley floors and changes of river channel pattern in the north polish plain during the late Wurm and Holocene. *Questiones Geographicae*, 4: 51–93.

Lane, E.W., 1955. The design of stable channels. *Transactions of the American Society of Civil Engineers*, 120: 1234–60.

Lane, S.N., Westaway, R.M. and Hicks, D.M., 2003. Estimation of erosion and deposition volumes in a large gravel-bed braided river using synoptic remote sensing. *Earth Surface Processes and Landforms*, 28: 249–71.

Langbein, W.B. and Leopold, L.B., 1966. *River Meanders: Theory of Minimum Variance*, United States Geological Survey Professional Paper 422H.

Langbein, W.B. and Schumm, S.A., 1958. Yield of sediment in relation to mean annual precipitation. *Transactions of the American Geophysical Union*, 39: 1076–84.

Large, A.R.G. and Petts, G.E., 1994. Rehabilitation of river margins. In: P. Calow and G.E. Petts (eds), *The Rivers Handbook*. Blackwell, Oxford, pp. 401–18.

Lawler, D.M., 1988. Environmental limits of needle ice: a global survey. *Arctic and Alpine Research*, 20: 137–59.

Lawless, M. and Robert, A., 2001. Scales of boundary resistance in coarse-grained channels. *Geomorphology* 39, 221–38.

Lee, J.G., Lovejoy, S.B. and Beasley, D.B., 1985. Soil loss reduction in Finley Creek, Indiana: an economic analysis of alternative policies. *Journal of Soil and Water Conservation*, 40(1): 132–5.

Leeder, M., 1999. *Sedimentology and Sedimentary Basins: From Turbulence to Tectonics*. Blackwell, Oxford.

Leopold, L.B. and Bull, W.B., 1979. Base level, aggradation, and grade. *Proceedings of the American Philisophical Society*, 123(3): 168–202.

Leopold, L.B. and Wolman, M.G., 1957. *River Channel Patterns: Braided, Meandering and Straight*. United States Geological Survey Professional Paper 282B.

Leopold, L.B., Emmett, W.W. and Myrick, R.M., 1966. *Channel and Hillslope Processes in a Semi-arid Area, New Mexico*. United States Geological Survey Professional Paper 323G.

Leopold, L.B., Wolman, M.G. and Miller, J.R., 1964. *Fluvial Processes in Geomorphology*. Freeman, San Francisco.

Lewin J. and Manton, M.M.M., 1975. Welsh floodplain studies: the nature of floodplain geometry. *Journal of Hydrology*, 25: 37–50.

Lewis, G.W. and Williams, G., 1984. *Rivers and Wildlife Handbook: A Guide to Practices which Further the Conservation of Wildlife on Rivers*, Royal Society for the Protection of Birds and Royal Society for Nature Conservation, London.

Lvovitch, M.I., Karasik, G.Y., Bratseva, N.I. and Medvedeva, G.P., 1991. *Contemporary Intensity of the World Land Intracontinental Erosion* USSR. Academy of Sciences, Moscow.

Macklin, M.G. and Lewin, J., 1989. Sediment transfer and transformation of an alluvial valley floor: the River South Tyne, Northumbria, UK. *Earth Surface Processes and Landforms*, 14: 233–46.

Macklin, M.G. and Lewin, J., 1993. Holocene river alluviation in Britain. *Zeitschrift für Geomorphologie, Supplement-Band* 88: 109–22.

Macklin, M.G. and Lewin, J., 1997. Channel, floodplain and drainage basin response to environmental change. In: C.R. Thorne, R.D. Hey and M.D. Newson (eds), *Applied Fluvial Geomorphology for River Engineering and Management*. John Wiley & Sons, Chichester, pp. 15–46.

MacVicar, B.J., 1999. '*Application of a Rational Model of Stream Equilibrium for Predicting Channel*

Adjustments'. M.A. Sci. thesis, University of British Columbia, Vancouver, BC.

Marion, D.A. and Weirach, F., 2003. Equal-mobility bed transport in a small step pool channel in the Ouachita Mountains. *Geomorphology*, 55: 139–54.

Markham, A.J. and Thorne, C.R., 1992. Geomorphology of gravel river bends. In: P. Billi, R.D. Hey, C.R. Thorne and P. Tacconi (eds), *Dynamics of Gravel-bed Rivers*. John Wiley & Sons, Chichester, pp. 433–50.

Meade, R.H., Dunne, T., Richey, J.E., Santos, U. d M. and Salati, E., 1985. Storage and remobilization of suspended sediment in the lower Amazon River of Brazil. *Science*, 228: 448–90.

Meyer-Peter, E. and Muller, R., 1948. Formulas for bedload transport, *International Association for Hydraulic Research Proceedings, Second Congress*, Stockholm, pp. 39–65.

Middleton, G.V. and Southard, J.B., 1977. *Mechanics of Sediment Movement*. Short Course 3 (Soc. Econ. Paleontologists and Mineralogists), Tulsa OK.

Middleton, G.V. and Southard, J.B., 1984. *Mechanics of Sediment Movement*. Society of Economic Paleontologists and Mineralogists (SEPM) Short Course 3.

Millar, R.G., 2000. Influence of bank vegetation on alluvial channel patterns. *Water Resources Research*, 36(4): 1109–18.

Miller, S.O., Ritter, D.F., Kochel, R.C. and Miller, J.R., 1993. Fluvial responses to land-use changes and climatic variations within the Dury Creek watershed, southern Illinois. *Geomorphology*, 6: 309–29.

Milliman, J.D. and Meade, R.H., 1983. World-wide delivery of river sediment to the oceans. *Journal of Geology*, 91: 1–21.

Milliman, J.D. and Syvitski, J.P.M., 1992. Geomorphic/tectonic control of sediment discharge to the ocean: the importance of small mountainous rivers. *Journal of Geology*, 100: 525–44.

Milliman, J.D., Yun-Shan, Q., Mei-E, R. and Saito, Y., 1987. Man's influence on the erosion and transport of sediment in Asian rivers: the Yellow River (Huanghe) example. *Journal of Geology*, 95: 751–62.

Morgan, R., 2005. *Soil Erosion and Conservation*. Blackwell, Oxford.

Morisawa, M., 1968. *Streams: Their Dynamics and Morphology*. McGraw-Hill, New York.

Morisawa, M., 1985. *Rivers*. Longman, Harlow.

Nanson, D.G., 1986. Episodes of vertical accretion and catastrophic stripping: a model of disequilibrium floodplain development. *Bulletin of the Geological Society of America*, 97: 1467–5.

Nanson, G.C. and Croke, J.C., 1992. A genetic classification of floodplains. *Geomorphology*, 4: 459–86.

Nanson, G.C. and Huang, H.Q., 1999. Anabranching rivers: divided efficiency leading to fluvial diversity. In: J.R. Miller and A. Gupta (eds), *Varieties of Fluvial Form*. John Wiley & Sons, Chichester, pp. 477–94.

Nanson, G. and Knighton, A.D., 1996. Anabranching rivers: their cause, character and classification. *Earth Surface Processes and Landforms*, 21: 217–39.

Nanson, G.C., Tooth, S. and Knighton, D., 2002. A global perspective on dryland rivers: perceptions, misconceptions and distinctions. In: L.J. Bull and M.J. Kirkby (eds), *Dryland Rivers: Hydrology and Geomorphology of Semi-arid Channels*. John Wiley & Sons, Chichester, pp. 17–54.

Nearing, L.J., Lane, L.J. and Lopes, V.L., 1994. Modelling soil erosion. In: R. Lal (ed.), *Soil Erosion: Research Methods*. St Lucie Press, Boca Raton FL, pp. 127–56.

Newson, M.D., Harper, D.M., Padmore, C.L., Kemp, J.L. and Vogel, B., 1998. A cost-effective approach for linking habitats, flow types and species requirements. *Aquatic Conservation: Marine and Freshwater Environments*, 8: 431–46.

Nicholas, A.P., 2005. Cellular modeling in fluvial geomorphology. *Earth Surface Processes and Landforms*, 30: 645–9.

Nicholas, A.P., Ashworth, P.J., Kirkby, M.J., Macklin, M.G. and Murray, T., 1995. Sediment slugs: large-scale fluctuations in fluvial transport rates and storage. *Progress in Physical Geography*, 19(4): 500–19.

O'Connor, J.E., Ely, L.L., Wohl, E.E., Stevens, L.E., Melis, T.S., Kale, V.S. and Baker, V.R., 1994. A 4500-year record of large floods on the Colorado River in the Grand Canyon, Arizona. *Journal of Geology*, 102(1): 1–10.

Parker, G., Klingeman, P.C. and McLean, D.G., 1982. Bedload and size distribution in paved gravel-bed streams. *Journal of the Hydraulic Division, ASCE* 108(4): 544–71.

Petersen, M.S., 1986. *River Engineering*. Prentice Hall, Englewood Cliffs NJ.

Petts, G.E., 1979. Complex response of river channel morphology subsequent to reservoir construction. *Progress in Physical Geography*, 3: 329–62.

Petts, G.E., 1984. *Impounded Rivers*. John Wiley & Sons, Chichester.

Petts, G.E. and Amoros, C., 1996. The fluvial hydrosystem. In: G.E. Petts and C. Amoros (eds), *Fluvial Hydrosystems*. Chapman & Hall, London, pp. 1–12.

Petts, G.E. and Lewin, J., 1979. Physical effects of reservoirs on river systems. In: G.E. Hollis (ed.), *Man's Impact on the Hydrological Cycle in the United Kingdom*. Geo Abstracts, Norwich, pp. 79–91.

Petts, G.E. and Maddock, I., 1996. Flow allocation for in-river needs. In: G.E. Petts and P. Calow (eds), *River Restoration: Selected Extracts from the Rivers Handbook*. Blackwell, Oxford, pp. 60–79.

Pickup, G., 1976. Alternative measures of river channel shape and their significance. *Journal of Hydrology (New Zealand)*, 15: 9–16.

Pickup, G. and Warner, R.F., 1976. Effects of hydrologic regime on magnitude and frequency of dominant discharge. *Journal of Hydrology*, 29: 51–75.

Pickup, G. and Reiger, W.A., 1979. A conceptual model of the relationship between channel characteristics and discharge. *Earth Surface Processes and Landforms*, 4: 37–42.

Pizzuto, J.E., 2003. Numerical modeling of alluvial landforms. In: G.M. Kondolf and H. Piégay (eds), *Tools in Fluvial Geomorphology*. John Wiley & Sons, Chichester, pp. 577–96.

Poesen, J., 1987. Transport of rock fragments by rill flow. A field study. *Catena Supplement*, 8: 35–54.

Poesen, J., Vandekerckhove, L., Nachtergaele, J., Oostwoud Wijdenes, D., Verstraeten, G. and van Wesemael, B., 2002. Gully erosion in dryland environments. In: L.J. Bull and M.J. Kirkby (eds), *Dryland Rivers: Hydrology and Geomorphology of Semi-arid Channels*. John Wiley & Sons, Chichester, pp. 229–62.

Prestegaard, K.L., 1983. Bar resistance in gravel bed streams at bankfull stage. *Water Resources Research*, 19: 472–6.

Prosser, I.P. and Williams, L., 1998. The effect of wildfire on runoff and erosion in native Eucalyptus forest. *Hydrological Processes*, 12: 251–65.

Przedwojski, B., Blazejewski, R. and Pilarczyk, K.W., 1995. River training techniques: fundamental design and applications. Balkema, Rotterdam.

Pye, K., 1994. Properties of sediment particles. In: K. Pye (ed.), *Sediment Transport and Depositional Processes*. Blackwell, Oxford, pp. 1–24.

Rajguru, S.N., Gupta, A., Kale, V.S., Mishra, S., Ganjoo, R.K., Ely, L.L., Enzel, Y. and Baker, V.R., 1995. Channel form and processes of the flood-dominated Narmada River, India. *Earth Surface Processes and Landforms*, 20: 407–21.

Reid, I., Bathurst, J.C., Carling, P.A., Walling, D.E. and Webb, B.W., 1997. Sediment erosion, transport and deposition. In: C.R. Thorne, R.D. Hey and M.D. Newson (eds), *Applied Fluvial Geomorphology for River Engineering and Management*. John Wiley & Sons, Chichester, pp. 95–135.

Rice, S. and Church, M., 1998. Grain size along two gravel-bed rivers: statistical variation, spatial pattern and sedimentary links. *Earth Surface Processes and Landforms*, 23: 345–63.

Richards, K.S., 1982. *Rivers: Form and Process in Alluvial Channels*. Methuen, London.

Richards, K.S., 2002. Drainage basin structure, sediment delivery and the response to environmental change. In: S.J. Jones and L.E. Frostick (eds), *Sediment Flux to Basins: Causes, Controls and Consequences*. Special Publications, 191. Geological Society, London, pp. 149–60.

Rinaldi, M. and Simon, A., 1998. Bed-level adjustments in the Arno River, central Italy. *Geomorphology*, 22: 57–71.

River Restoration Centre, 1999. *River Restoration: Manual of Techniques: Restoring the River Cole and River Skerne, UK*. River Restoration Centre, Silsoe.

Robert, A., 2003. *River Processes: An Introduction to Fluvial Dynamics*. Arnold, London.

Roberts, C.R., 1989. Flood frequency and urban-induced channel change: some British examples. In: K. Beven and P.A. Carling (eds), *Floods: Hydrological, Sedimentological and Geomorphological Implications*. John Wiley & Sons, Chichester, pp. 57–82.

Rodnight, H., Duller, G.A.T., Wintle, A.G. and Tooth, S., 2006. Assessing the reproducibility and accuracy of optical dating of fluvial deposits. *Quaternary Geochronology*, 1: 109–20.

Rotnicki, K., 1991. Retrodiction of palaeodischarges of meandering and sinuous alluvial rivers and its palaeohydroclimatic implications. In: L. Starkel, K.J. Gregory and J.B. Thornes (eds), *Temperate Palaeohydrology*. John Wiley & Sons, Chichester, pp. 431–71.

Rosgen, D.L., 1994. A classification of natural rivers. *Catena*, 22: 169–99.

Rowntree, K. and Wadeson, R.A., 1996. Translating channel geomorphology into aquatic habitat: application of the hydraulic biotope concept to assessment of discharge related habitat changes. In M. Leclerc (ed.), *Ecohydraulics 2000: Proceedings of the Second International Symposium of Habitat Hydraulics*. International Association of Hydraulics Research Quebec, pp. 342–51.

Rumsby, B.T. and Macklin, M.G., 1994. Channel and floodplain response to recent abrupt climate change: the Tyne basin, northern England. *Earth Surface Processes and Landforms*, 19: 499–515.

Said, R., 1993. *The River Nile: Geology, Hydrology and Utilization*. Pergamon Press, Oxford.

Schaller, M., Hovius, N., Willett, S.D., Ivy-Ochs, S., Synal, H.-A. and Chen, M.-C., 2005. Fluvial bedrock incision in the active mountain belt of Taiwan from *in situ*-produced cosmogenic nuclides. *Earth Surface Processes and Landforms*, 30: 955–71.

Schlichting, H, 1979. *Boundary Layer Theory*. Seventh edition. McGraw-Hill, New York.

Schumm, S.A., 1968. *River Adjustment to altered Hydrologic Regimen: Murrumbidgee River and Palaeochannels, Australia*. United States Geological Survey Professional Paper 598.

Schumm, S.A., 1969. River metamorphosis. *Journal of the Hydraulic Division, ASCE*, 95(HY1): 255–73.

Schumm, S.A., 1977. *The Fluvial System*. John Wiley & Sons, New York.

Schumm, S.A., 1979. Geomorphic thresholds: the concept and its applications. *Proceedings of the Institute of British Geographers* 4: 485–515.

Schumm, S.A., 1988. Variability of the fluvial system in space and time. In: T. Rosswall, R.G. Woodmansee and P.G. Risser (eds), *Scales and Global Change: Spatial and Temporal Variability in Biospheric and Geospheric Processes*. SCOPE: 35. John Wiley & Sons, Chichester, pp. 225–50.

Schumm, S.A. and Khan, H.R., 1972. Experimental study of channel patterns. *Geological Society of America Bulletin*, 83, 1755–70.

Schumm, S.A. and Lichty, R.W., 1963. *Channel Widening and Flood-plain Construction along Cimarron River in South-western Kansas*. United States Geological Survey Professional Paper 352D.

Sear, D.A., Darby, S.E., Thorne, C.R. and Brookes, A., 1995. Sediment-related river maintenance: the role of fluvial geomorphology. *Earth Surface Processes and Landforms*, 20: 629–47.

Shields, A., 1936. *Anwendung der Ähnlichkeitsmechanic und Turbulenzforschung auf die Geschiebebewegung*. Report 26, *Mitteil Preuss. Versuchsant*. Wasserbau und Schiffsbau, Berlin.

Shiono, K. and Knight, D.W., 1991. Turbulent open-channel flows with variable depth across the channel *Journal of Fluid Mechanics*, 222: 617–46.

Simon, A. and Castro, J., 2003. Measurement and analysis of alluvial channel form. In: G.M. Kondolf and H. Piégay (eds), *Tools in Fluvial Geomorphology*. John Wiley & Sons, Chichester: 291–322.

Simons, D.B. and Richardson, E.V., 1962. Resistance to flow in alluvial channels. *Transactions of the American Geophysical Union*, 127: 927–53.

Simons, D.B. and Richardson, E.V., 1966. *Resistance to Flow in Alluvial Channels*. United States Geological Survey Professional Paper 422–J.

Smart, C.C. and Brown, M.C., 1981. Some results and limitations in the application of hydraulic geometry to vadose stream passages, Proceedings of the Eighth International Congress on Speleology, Kentucky, pp. 724–5.

Smith, N.D., 1974. Sedimentology and bar formation in the upper Kicking Horse River, a braided outwash stream. *Journal of Geology*, 82: 205–23.

Soil Science Society of America, 1996. *Glossary of Soil Science Terms*. Soil Science Society of America, Madison, WI.

Starkel, L., 1987. Man as a cause of sedimentologic changes in the Holocene. *Striae*, 26: 5–12.

Starkel, L., 1991a. The Vistula River valley: a case study for Central Europe. In: L. Starkel, K.J. Gregory and J.B. Thornes (eds), *Temperate Palaeohydrology*. John Wiley & Sons, Chichester, pp. 171–88.

Starkel, L., 1991b. Long-distance correlation of fluvial events in the temperate zone. In: L. Starkel, K.J. Gregory and J.B. Thornes (eds), *Temperate Palaeohydrology*. John Wiley & Sons, Chichester, pp. 471–95.

Stedinger, J.R., and Baker, V.R., 1987. Surface water hydrology: historical and paleoflood information. *Reviews of Geophysics*, 25:119–24.

Stevens, M.A., Simons, D.B. and Richardson, E.V., 1975. Nonequilibrium river form. *Journal of the Hydraulics Division, ASCE*, 101 (HY5): 557–67.

Summerfield, M.A., 1991. *Global Geomorphology: An Introduction to the Study of Landforms*. Longman, Harlow.

Tabata, K.K. and Hickin, E.J., 2003. Interchannel hydraulic geometry and hydraulic efficiency of the anastomosing Columbia River, southeastern British Columbia, Canada. *Earth Surface Processes and Landforms*, 28(8): 837–52.

Thompson, D.M., 2001. Random controls on semi-rhythmic spacing of pools and riffles in construction-dominated rivers. *Earth Surface Processes and Landforms*, 26: 1195–212.

Thorne, C.R., 1997. Channel Types and Morphological Classification. In: C.R. Thorne, R.D. Hey and M.D. Newson, 1997. *Applied Fluvial Geomorphology for River Engineering and Management*. John Wiley & Sons, Chichester, pp. 175–222.

Thorne, C.R., 1998. *Stream Reconnaissance Handbook: Geomorphological Investigation and Analysis of River Channels*. John Wiley & Sons, Chichester.

Thorne, C.R. and Osman, A.M., 1988. Riverbank stability analysis II: applications. *Journal of the Hydraulic Division, Proceedings of the American Society of Civil Engineers*, 114: 151–72.

Thorp, M. and Thomas, M., 1992. The timing of alluvial sedimentation and floodplain formation in the lowland humid tropics of Ghana, Sierra Leone and western Kalimantan (Indonesian Borneo). *Geomorphology*, 4: 409–22.

Tinkler, K.J., 1971. Active valley meanders in south-central Texas and their wider significance. *Geological Society of America Bulletin*, 82: 1783–800.

Tinkler, K.J., 1993. Fluvially sculpted bedforms in Twenty Mile Creek, Niagara Peninsula, Ontario. *Canadian Journal of Earth Science*, 30: 945–53.

Tinkler, K.J. and Wohl, E.E., 1998. A primer on bedrock channels. In: K.J. Tinkler and E.E. Wohl (eds), *Rivers over Rock: Fluvial Processes in Bedrock Channels*. American Geophysical Union Geophysical Monograph 107, Washington DC, pp. 1–18.

Tooth, S., 1999. Floodouts in central Australia. In: A.J. Miller and A. Gupta (eds), *Varieties of Fluvial Form*. John Wiley & Sons, Chichester, pp. 219–47.

Tooth, S., 2000. Downstream changes in dryland river channels: the northern plains of arid central Australia. *Geomorphology*, 34: 33–54.

Tooth, S. and McCarthy, T.S., 2004. Anabranching in mixed bedrock-alluvial rivers: the example of the Orange River above Augrabies Falls, Northern Cape Province, South Africa. *Geomorphology*, 57: 235–62.

Tooth, S. and Nanson, G.C., 1999. Anabranching rivers on the northern plains of arid central Australia. *Geomorphology*, 29: 211–33.

Trimble, W.S., 1983. A sediment budget for Coon Creek Basin in the driftless area, Wisconsin. *American Journal of Science*, 283: 454–74.

Troeh, F.R., Hobbs, J.A. and Donahue, R.L., 2003. *Soil and Water Conservation for Productivity and Environmental Protection*. Prentice Hall, Englewood Cliffs NJ.

Urban, M.A., 2002. Conceptualizing anthropogenic change in fluvial systems drainage development on the Upper Embarras River, Illinois. *Professional Geographer*, 54(2): 204–18.

van Niekerk, A.W., Heritage, G.L., Broadhurst, L.J. and Moon, B.P., 1999. Bedrock anastomosing channel systems: morphology and dynamics in the Sabie River, Mpumalanga Province, South Africa. In: A.J. Miller and A. Gupta (eds), *Varieties of Fluvial Form*. John Wiley & Sons, Chichester, pp. 53–79.

Vericat, D. and Batella, R.J., 2005. Sediment transport in a highly regulated fluvial system during two consecutive floods (lower Ebro River, north-east Iberian peninsula). *Earth Surface Processes and Landforms*, 30: 385–402.

Wainwright, J. and Mulligan, M., 2004. Modelling and model building. In: J. Wainwright and M. Mulligan (eds) *Environmental Modelling: Finding Simplicity in Complexity*. John Wiley & Sons. Chichester, pp. 7–73.

Walling, D.E. and Webb, B.W., 1983. Patterns of sediment yield. In: K.J. Gregory (ed.), *Background to Palaeohydrology*. John Wiley & Sons, Chichester, pp. 69–100.

Ward, R.C., 1984. On the response of precipitation of headwater streams in humid areas. *Journal of Hydrology*, 74: 171–89.

Ward, R.C. and Robinson, M., 1990. *Principles of Hydrology*. McGraw-Hill, London.

WCD, 2000. *Dams and Development: A New Framework for Decision-making. The Report of the World Commission on Dams*, Earthscan Publications, London and Sterling VA.

Wende, R. and Nanson, G., 1998. Anabranching rivers: ridge-forming alluvial channels in tropical northern Australia. *Geomorphology*, 22: 205–24.

Wilcock, P.R., 1992. Experimental investigation of the effect of mixture properties on transport dynamics. In: P. Billi, R.D. Hey, C.R. Thorne and P. Tacconi (eds), *Dynamics of Gravel-bed Rivers*. John Wiley & Sons, Chichester, pp. 109–39.

Willgoose, G., Bras, I. and Rodriquez-Iturbe, I., 1991. Results from a new model of river basin evolution. *Water Resources Research*, 30(7): 2261–85.

Williams, G.P., 1978. Bankfull discharge of rivers. *Water Resources Research*, 14: 1141–58.

Williams, G.P. and Wolman, M.G., 1984. *Downstream Effects of Dams on Alluvial Rivers*. United States Geological Survey Professional Paper 1286.

Wilson, E.M., 1990. *Engineering Hydrology*. Fourth edition. Macmillan, Basingstoke.

Wohl, E.E., 1998. Bedrock channel morphology in relation to erosional processes. In: K.J. Tinkler and E.E. Wohl (eds), *Rivers over Rock: Fluvial Processes in Bedrock Channels*. Geophysical Monograph Series. American Geophysical Union, Washington DC, pp. 133–51.

Wolman, M.G., 1967. A cycle of sedimentation and erosion in urban river channels. *Geografiska Annaler*, 49A: 385–95.

Wolman, M.G. and Miller, J.P., 1960. *Magnitude and Frequency of Forces in Geomorphic Processes*, United States Geological Survey Professional Paper 282C.

Woodward, J. and Foster, I.D.L., 1997. Erosion and suspended sediment transfer in river catchments. *Geography*, 82(4): 353–76.

Xu, J., 1997. Study of sedimentation zones in a large sand-bed braided river: an example from the Hanjiang River of China. *Geomorphology*, 21: 152–65.

Zaslavsky, D. and Sinai, G., 1981. Surface hydrology, I. Explanation of phenomena. *Journal of the Hydraulic Division, Proceedings of the American Society of Civil Engineers*, 107: 1–16.

Zhong, Y. and Power, G., 1996. Environmental impacts of hydroelectric projects on fish resources in China. *Regulated Rivers: Research and Management*, 12: 81–98.

INDEX

eBooks

eBooks – at www.eBookstore.tandf.co.uk

A library at your fingertips!

eBooks are electronic versions of printed books. You can store them on your PC/laptop or browse them online.

They have advantages for anyone needing rapid access to a wide variety of published, copyright information.

eBooks can help your research by enabling you to bookmark chapters, annotate text and use instant searches to find specific words or phrases. Several eBook files would fit on even a small laptop or PDA.

NEW: Save money by eSubscribing: cheap, online access to any eBook for as long as you need it.

Annual subscription packages

We now offer special low-cost bulk subscriptions to packages of eBooks in certain subject areas. These are available to libraries or to individuals.

For more information please contact
webmaster.ebooks@tandf.co.uk

We're continually developing the eBook concept, so keep up to date by visiting the website.

www.eBookstore.tandf.co.uk